BROWNING AUTOMATIC RIFLE, CALIBER .30, M1918 WITHOUT BIPOD

FIELD MANUAL

BY WAR DEPARTMENT

DISCLAIMER:

This manual is sold for historic research purposes
only, as an entertainment. It contains obsolete
information and is not intended to be used as part
of an actual operation or maintenance training
program. No book can substitute for proper training
by an authorized instructor.

FM 23–20

BASIC FIELD MANUAL

BROWNING AUTOMATIC RIFLE
CALIBER .30, M1918
WITHOUT BIPOD

Prepared under direction of the
Chief of Infantry

UNITED STATES
GOVERNMENT PRINTING OFFICE
WASHINGTON: 1940

WAR DEPARTMENT,
Washington, *October 1, 1940*.

FM 23–20, Basic Field Manual, Browning Automatic Rifle, Caliber .30, M1918, without Bipod, is published for the information and guidance of all concerned.

[A. G. 062.11 (5–11–40).]

By order of the Secretary of War:

G. C. MARSHALL,
Chief of Staff.

Official:

E. S. ADAMS,
Major General,
The Adjutant General.

TABLE OF CONTENTS

		Paragraphs	Pages
CHAPTER 1. MECHANICAL TRAINING.			
Section I.	General	1–3	1
II.	Disassembly and assembly	4–12	4
III.	Care and cleaning of the rifle	13–15	36
IV.	Functioning	16–24	39
V.	Operation	25–34	55
VI.	Immediate action and stoppages	35–38	58
VII.	Spare parts and accessories	39–40	64
VIII.	Ammunition	41–48	61
CHAPTER 2. MARKSMANSHIP—KNOWN-DISTANCE TARGETS.			
Section I.	General	49–51	71
II.	Preparatory marksmanship training	52–75	72
III.	Courses to be fired	76–78	100
IV.	Range practice	79–88	105
V.	Regulations governing record practice	89–119	113
VI.	Targets and ranges	120–121	123
CHAPTER 3. MARKSMANSHIP—MOVING GROUND TARGETS.			
Section I.	General	122–123	128
II.	Moving vehicles	124–126	128
III.	Moving personnel	127–128	130
IV.	Moving targets and ranges and range precautions	129–130	130
CHAPTER 4. MARKSMANSHIP—AIR TARGETS.			
Section I.	Nature of air targets for the automatic rifle	131–132	134
II.	Technique of antiaircraft fire	133–137	134
III.	Marksmanship training	138–142	136
IV.	Miniature range practice	143–146	144
V.	Towed-target firing	147–151	147
VI.	Ranges, targets, and equipment	152–157	150
CHAPTER 5. TECHNIQUE OF FIRE.			
Section I.	Introduction	158–160	162
II.	Range estimation	161–165	163
III.	Target designation	166–173	166
IV.	Rifle fire and its effect	174–180	174
V.	Application of fire	181–188	177
VI.	Landscape-target firing	189–196	180
VII.	Firing at field targets	197–202	187
CHAPTER 6. ADVICE TO INSTRUCTORS.			
Section I.	General	203	192
II.	Mechanical training	204	192
III.	Marksmanship—known-distance targets	205–219	193
IV.	Marksmanship—air targets	220–224	207
V.	Technique of fire	225–233	213
INDEX			219

III

BASIC FIELD MANUAL

BROWNING AUTOMATIC RIFLE, CALIBER .30, M1918, WITHOUT BIPOD

(This manual supersedes chapter 2, part one, Basic Field Manual, Volume III, March 25, 1932, and TR 1300–30E, October 12, 1939)

CHAPTER 1

MECHANICAL TRAINING

Paragraphs

SECTION I. General _____ 1–3
  II. Disassembly and assembly_____ 4–12
  III. Care and cleaning of the rifle_____ 13–15
  IV. Functioning_____ 16–24
  V. Operation _____ 25–34
  VI. Immediate action and stoppages_____ 35–38
  VII. Spare parts and accessories_____ 39–40
  VIII. Ammunition_____ 41–48

SECTION I

GENERAL

■ 1. OBJECT.—This chapter is designed to give the soldier training that will insure his ability to maintain the rifle and keep it in operation.

■ 2. DESCRIPTION OF RIFLE.—The Browning automatic rifle, caliber .30, M1918, without bipod, is an air-cooled, gas-operated, magazine-fed, shoulder weapon. (See fig. 1.) It weighs 15 pounds 14 ounces. The ammunition is loaded in magazines of 20 rounds each. The weight of the magazine when empty is 7 ounces; when filled, 1 pound 7 ounces. The design permits semiautomatic and automatic fire.

■ 3. FIREPOWER.—This rifle is capable of semiautomatic fire at the rate of 100 rounds per minute. Its rate of effective sustained fire is about 40 rounds per minute.

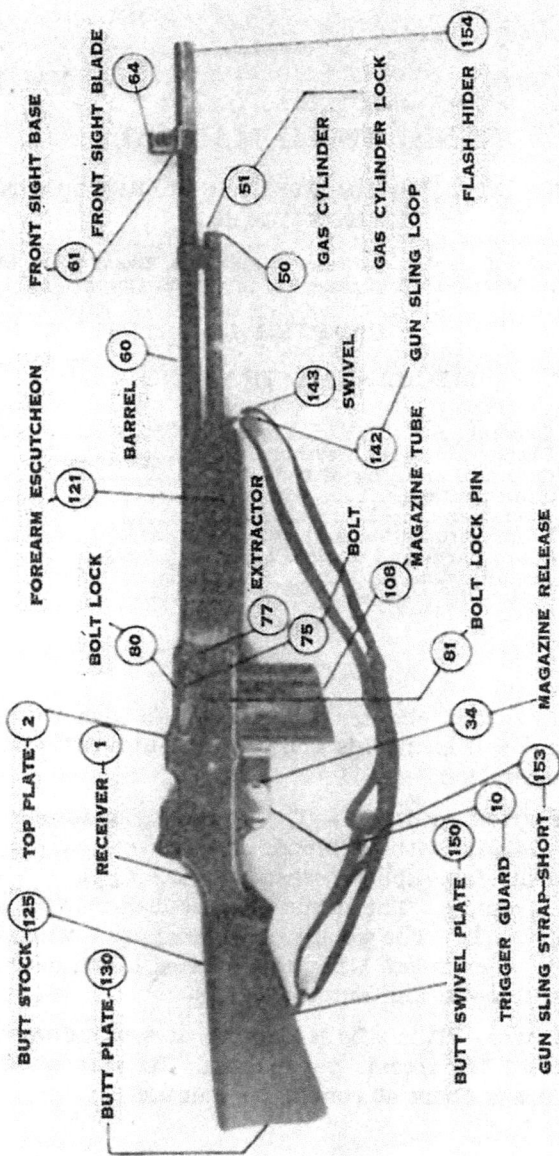

FRONT SIGHT BASE

FRONT SIGHT BLADE — 64

GAS CYLINDER

GAS CYLINDER LOCK

GUN SLING LOOP

FLASH HIDER — 154

61

51

50

FOREARM ESCUTCHEON

BARREL — 60

143

SWIVEL

142

MAGAZINE TUBE

121

EXTRACTOR

BOLT

MAGAZINE LOCK PIN

BOLT LOCK

77

108

80

75

BOLT LOCK PIN

81

MAGAZINE RELEASE

TOP PLATE — 2

RECEIVER — 1

34

153

BUTT STOCK — 125

150

10

BUTT SWIVEL PLATE

TRIGGER GUARD

GUN SLING STRAP — SHORT

BUTT PLATE — 130

① Right side view.

2

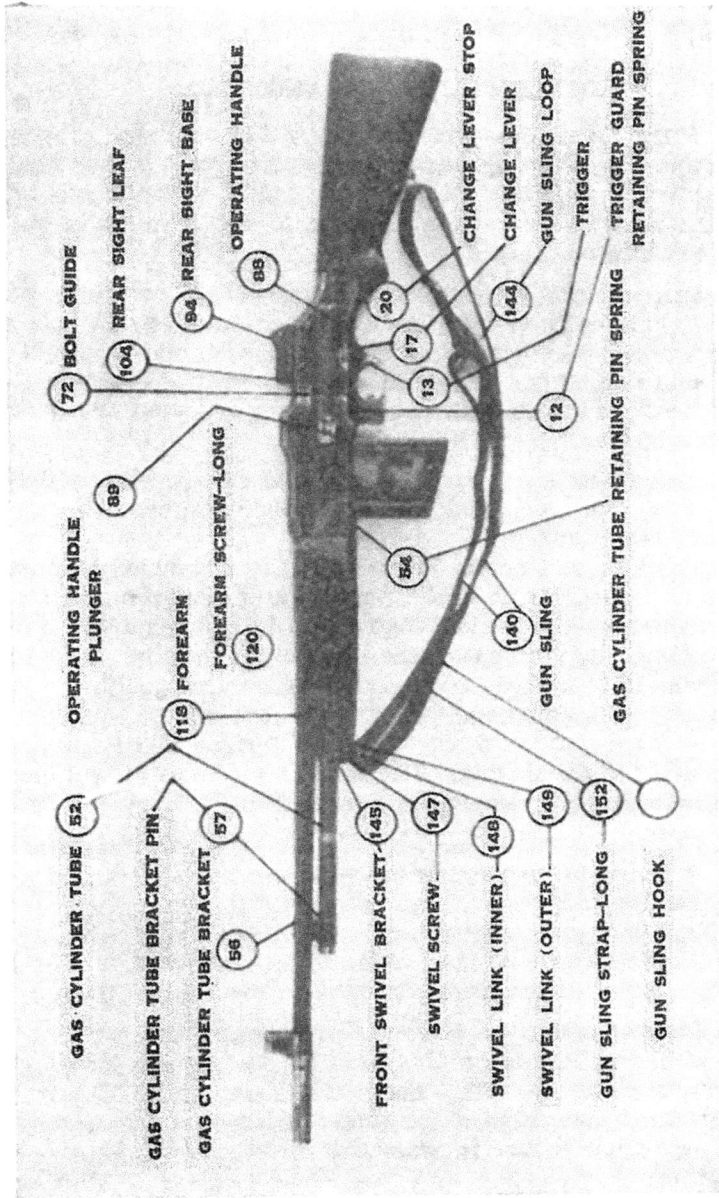

72 BOLT GUIDE

REAR SIGHT LEAF

94 REAR SIGHT BASE

OPERATING HANDLE

88

104

CHANGE LEVER STOP

20

CHANGE LEVER

17

144 GUN SLING LOOP

13

TRIGGER

12

TRIGGER GUARD
RETAINING PIN SPRING

OPERATING HANDLE
PLUNGER

89

118 FOREARM

FOREARM SCREW—LONG

120

GAS CYLINDER TUBE 52

GAS CYLINDER TUBE BRACKET PIN

GAS CYLINDER TUBE BRACKET

57

56

54

140

GUN SLING

GAS CYLINDER TUBE RETAINING PIN SPRING

FRONT SWIVEL BRACKET 145

SWIVEL SCREW 147

SWIVEL LINK (INNER) 148

SWIVEL LINK (OUTER) 149

GUN SLING STRAP—LONG 152

GUN SLING HOOK

① Left side view.

FIGURE 1.—Browning automatic rifle, caliber .30. M1918, without bipod.

3

SECTION II

DISASSEMBLY AND ASSEMBLY

■ 4. WHEN TAKEN UP.—This training will be taken up as soon as practicable after the soldier receives his rifle. In any case it will be completed before any firing is done with the rifle by the individual. Instruction in the care and cleaning of the rifle will also be covered.

■ 5. ORGANIZATION.—In units the size of a company or platoon, all enlisted men are combined into one or more groups under their officers, or selected noncommissioned officers, as instructors. Other noncommissioned officers supervise the work as directed. Squad leaders supervise the work of their squads.

■ 6. CARE TO BE EXERCISED.—*a.* The rifle can be readily disassembled and assembled without applying force. The use of force is prohibited.

b. The rifle will not be disassembled or assembled against time as this serves no useful purpose and results in burring and damaging the parts. Instruction in disassembling and assembling the automatic rifle blindfolded may be given to men who have passed their tests in mechanical training. In all work in disassembling the rifle, the men will be taught to lay the parts out on a smooth, clean surface in the proper sequence for assembling. The trigger mechanism will not be disassembled or assembled blindfolded.

■ 7. NOMENCLATURE.—The names of the parts to which reference is made in mechanical training are readily learned as this training progresses. Instructors will therefore take care to name the parts clearly and correctly in their work. A sufficient knowledge of the nomenclature of the rifle is gained by the soldier during the instruction in mechanical training.

■ 8. DISASSEMBLING.—*a. General.*—Authorized disassembly by the soldier is limited to that required for proper care and maintenance of the rifle. Further disassembly will be done under the supervision of an officer or ordnance personnel. The individual soldier is prohibited from disassembling the following:

4

(1) Forearm group.

(2) Barrel group.

(3) Butt stock and buffer group.

(4) Rear-sight group.

(5) Receiver group.

b. Sequence.—The disassembly of the rifle authorized to be performed by the individual soldier without supervision is performed in the following sequence:

(1) *Operating group.*

(*a*) Cock the rifle.

(*b*) Remove gas cylinder tube retaining pin.

(*c*) Remove gas cylinder tube and forearm (let mechanism forward easily).

(*d*) Remove trigger guard retaining pin.

(*e*) Remove trigger guard.

(*f*) Remove recoil spring guide and recoil spring.

(*g*) Push hammer pin through hammer pin hole in receiver.

(*h*) Remove operating handle.

(*i*) Remove hammer pin.

(*j*) Remove hammer.

(*k*) Remove slide.

(*l*) Push out bolt guide.

(*m*) Remove bolt, bolt lock, and bolt link.

(*n*) Remove firing pin.

(*o*) Remove bolt link pin and bolt link.

(*p*) Remove extractor and spring.

(2) *Trigger mechanism.*

(*a*) Remove ejector.

(*b*) Remove magazine catch spring.

(*c*) Remove magazine catch pin.

(*d*) Remove magazine catch.

(*e*) Remove magazine release.

(*f*) Remove sear spring.

(*g*) Remove trigger pin.

(*h*) Remove trigger and connector.

(*i*) Remove sear pin.

(*j*) Remove sear.

(*k*) Remove sear carrier and counter-recoil spring.

(*l*) Remove change lever spring.

(*m*) Remove change lever.

c. Method.—The following detailed explanation of the method of disassembling the automatic rifle is furnished as an aid to instructors:

(1) *Operating group.*—Lay the rifle on the table, barrel down, pointing to the left. Cock the piece. This must be done in order that the gas cylinder tube may clear the gas piston and gas cylinder tube bracket. Turn the cylinder tube retaining pin spring ((54) fig. 1 ②), 180° in a clockwise direction and lift out gas cylinder tube (52) retaining pin. Remove the gas cylinder tube and forearm (118). Let the slide (fig. 2) forward *easily* by pressing the trigger with the thumb of the right hand and at the same time grasp the slide with the left hand so that the middle and index fingers are astride the gas piston. Turn the trigger guard retaining pin spring ((12), fig. 1 ②), 90° in a clockwise direction and lift out the pin. Lift out the trigger guard group.

Remove the recoil-spring guide (Fig. 2) and the recoil spring by pressing the right index finger on the checkered surface of the recoil-spring rod and turning it until the ends are clear of the retaining shoulders. Line up the hammer-pin holes on the receiver and the operating handle by inserting the point of the recoil-spring guide or the point of a dummy cartridge in the hole on the operating handle with the right hand, press against the hammer pin and push the operating handle backward with the left hand. The recoil-spring guide will push the hammer pin through its hole in the receiver as the hammer pin registers with the latter. Remove the operating handle ((88), fig. 1 ②) by pulling it straight to the rear. Remove the hammer pin. Push the hammer (fig. 2) forward out of its seat in the slide and lift it out of the receiver. Remove the slide by pulling it forward out of the receiver, being careful that the bolt link is pushed well down, thus allowing the slide to clear. In removing the slide, take care to avoid striking the gas piston or rings against the gas cylinder tube bracket ((56), fig. 1 ②). Force the bolt guide (72) out with the left thumb or the point

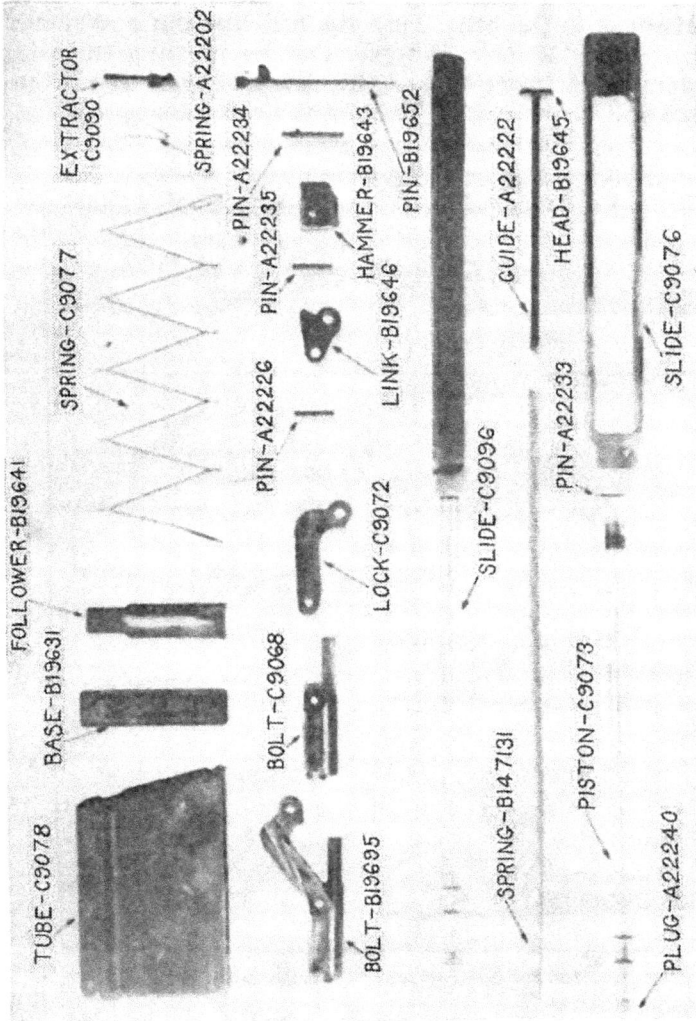

FIGURE 2.—Gas piston and bolt group parts.

of a bullet.   Lift out the bolt, bolt lock, and bolt link by pulling them slowly to the rear end of the receiver and up with right thumb and forefinger.   Pull out the firing pin (fig. 2) from its way in the bolt.   Push the bolt-link pin and remove the bolt link.   Remove the extractor by pressing the point of a dummy cartridge against the claw and exerting pressure upward and to the front.   Remove the extractor spring.

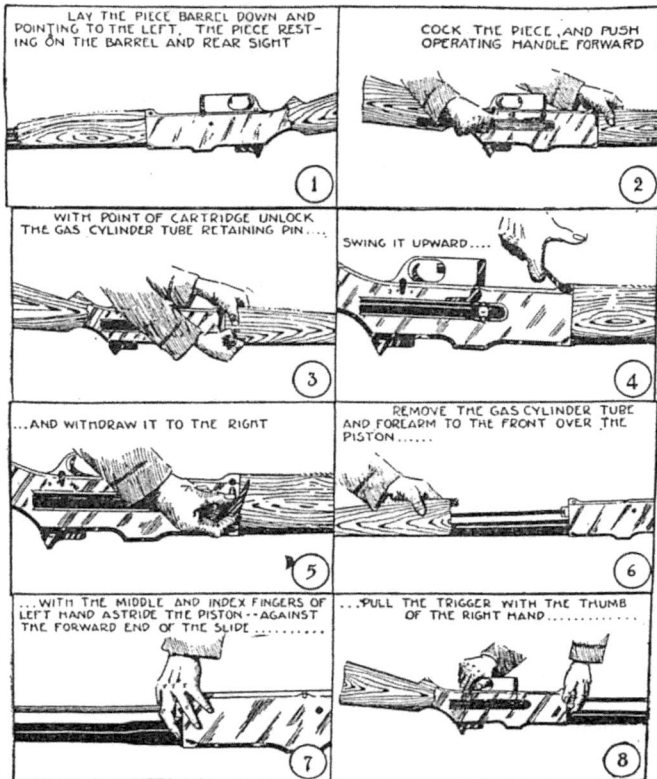

FIGURE 3.—Method of disassembling.

(2) *Trigger mechanism.*—(a) Depress the ejector lock with the point of a dummy cartridge. Hold the thumb in front of the magazine-catch spring to prevent it from flying out and then slide the ejector out of its seat. (See fig. 5.) Remove the magazine-catch spring. Remove the magazine-catch pin, lift out the magazine catch (see fig. 6) and magazine release (see fig. 7).

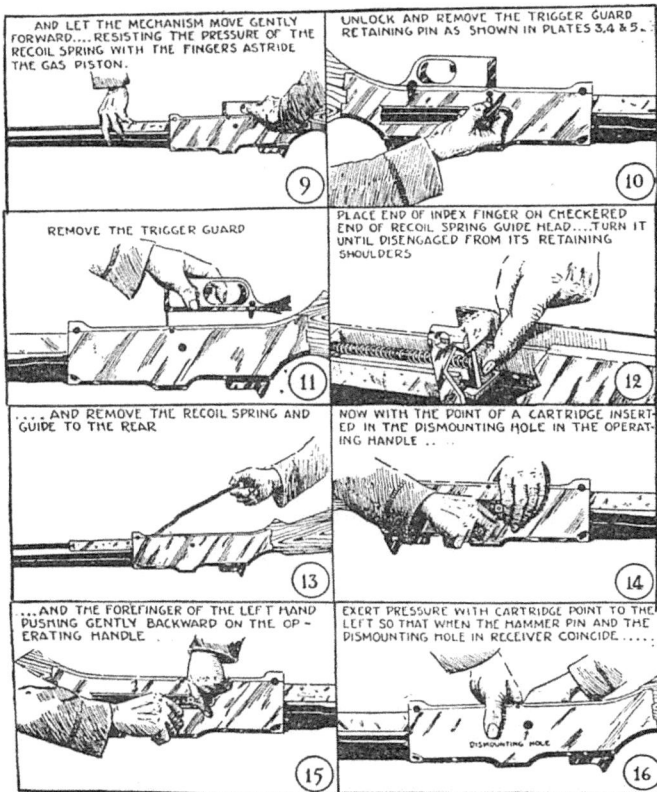

FIGURE 3.—Method of disassembling—Continued.

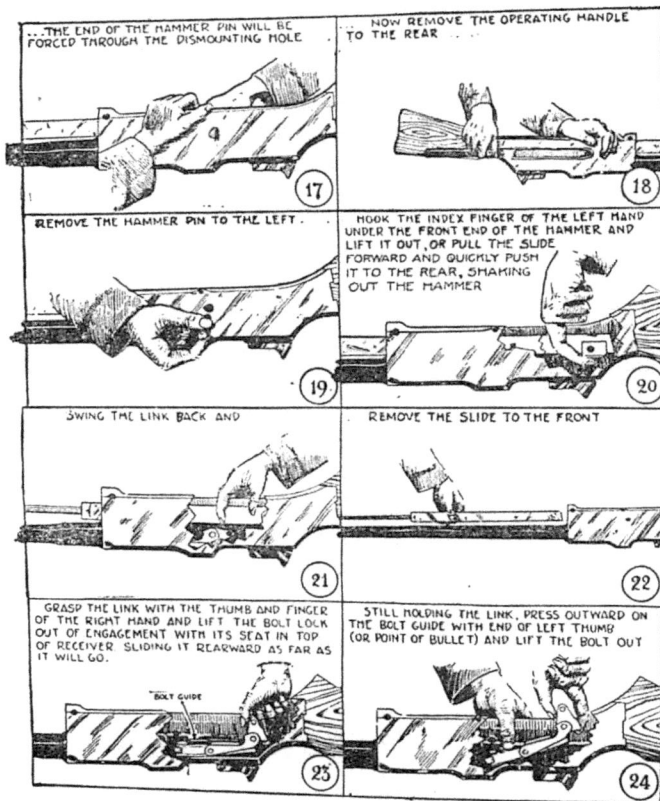

FIGURE 3.—Method of disassembling—Continued.

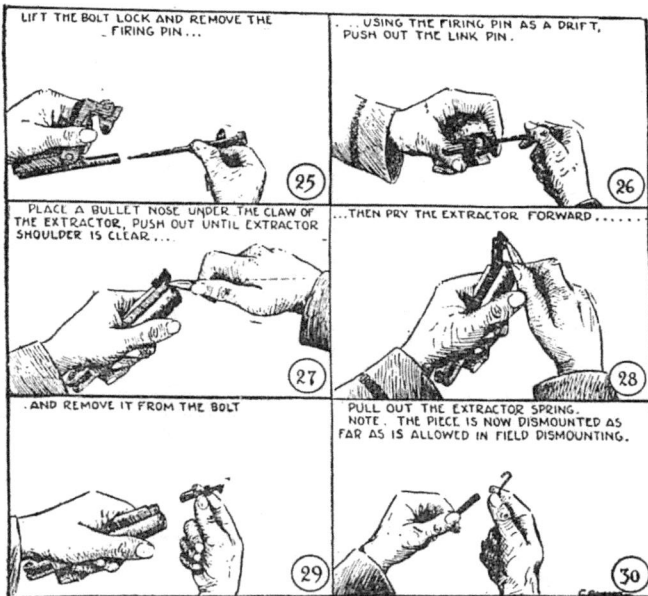

FIGURE 3.—Method of disassembling—Continued.

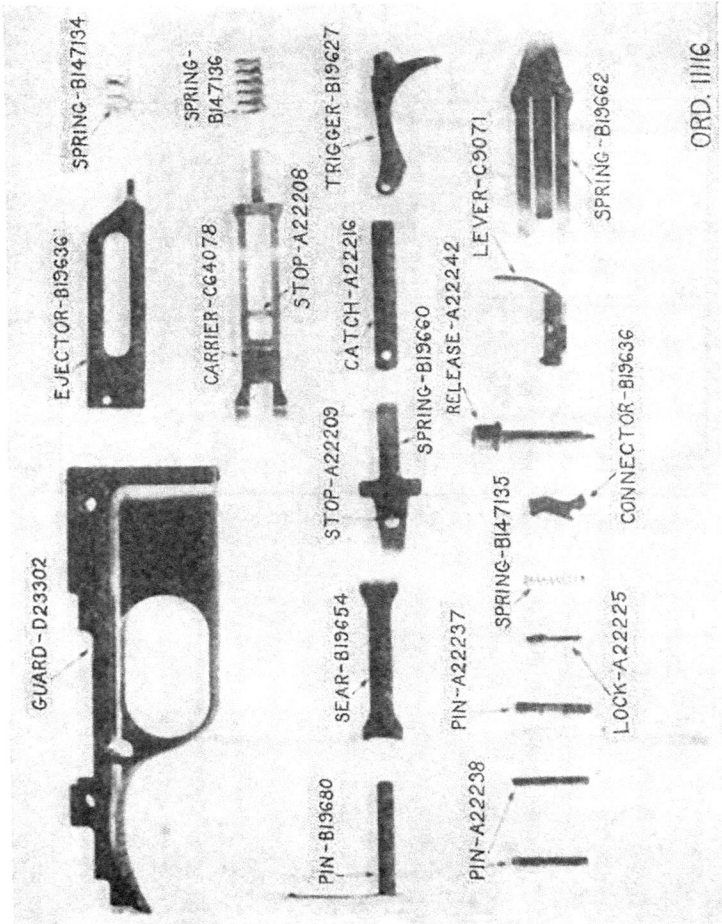

FIGURE 4.—Trigger guard group.

(b) Insert the trigger guard retaining pin spring under the sear spring above the connector stop. Pry up, pressing against the sear spring with thumb and pull it out to the rear. (See fig. 8.) Push out the trigger pin (see fig. 9). The trigger pin must always be removed before the sear pin in order that the tension of the counterrecoil spring will always be on the sear pin. Remove the trigger and connector through top of trigger guard. (See fig. 10.) Push out the sear pin, using the recoil spring guide (fig. 2). Remove the sear (see fig. 11). Pry up on the sear carrier (see fig. 12) and lift out the sear carrier and counterrecoil spring. Remove the change lever spring by prying the bent end out of its seat with the rounded end of the sear spring and moving the change lever from front to rear. When clear of the change lever, push it out the rest of the way by pressing with the thumb against the sear stop. Pull out the change lever.

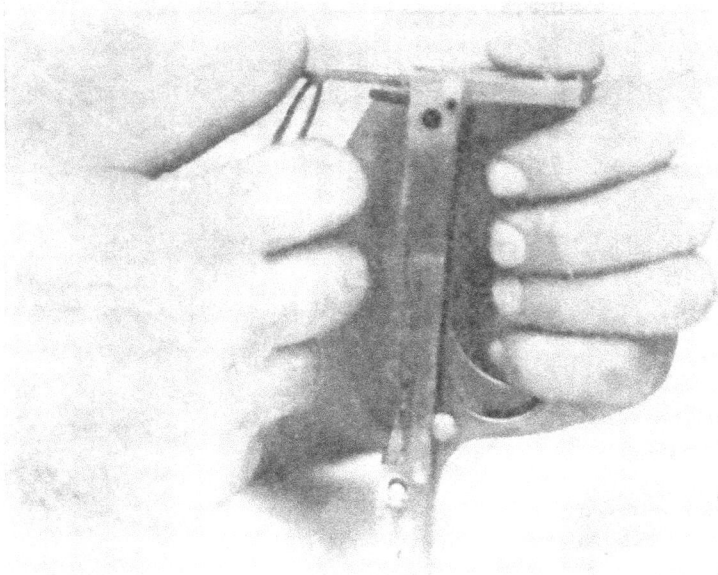

FIGURE 5.—Removing ejector.

13

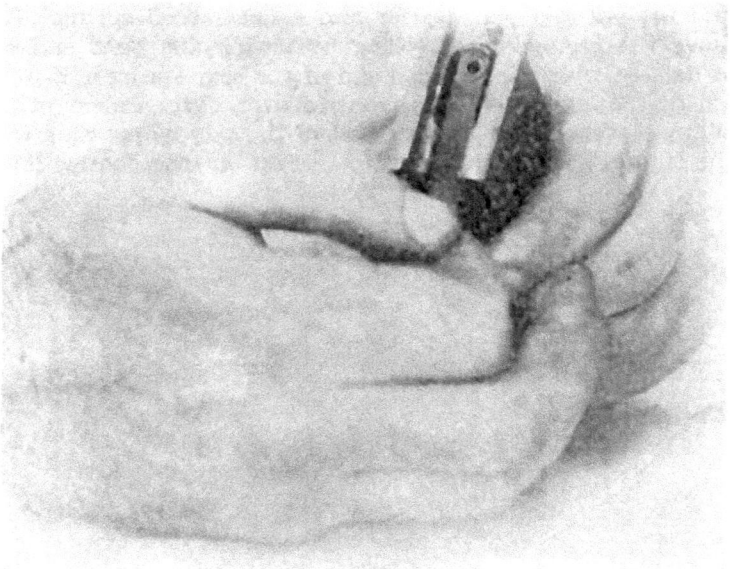

FIGURE 6.—Removing magazine catch.

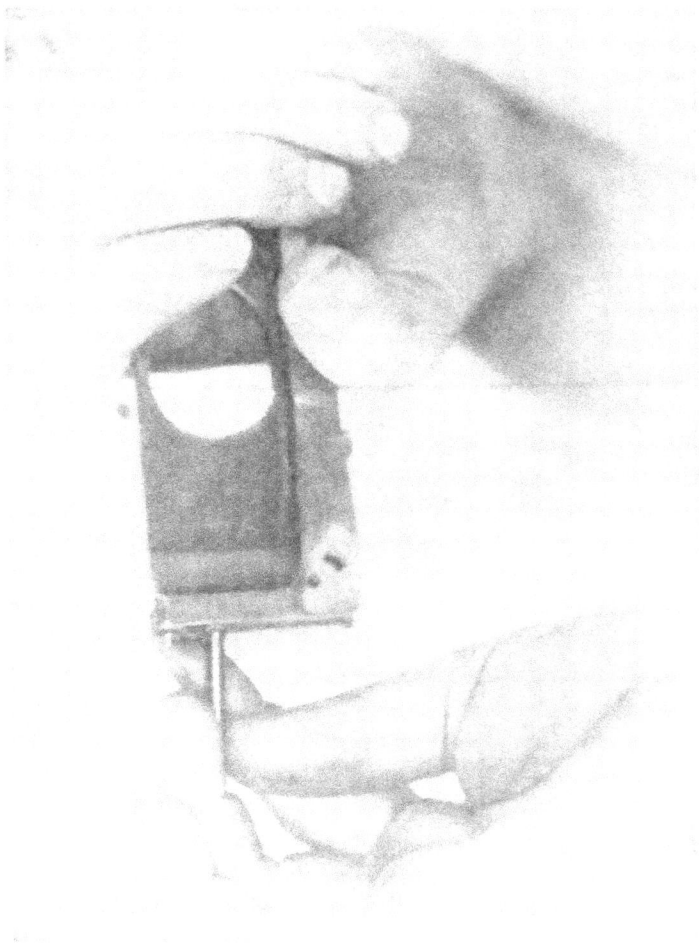

FIGURE 7.—Removing magazine release.

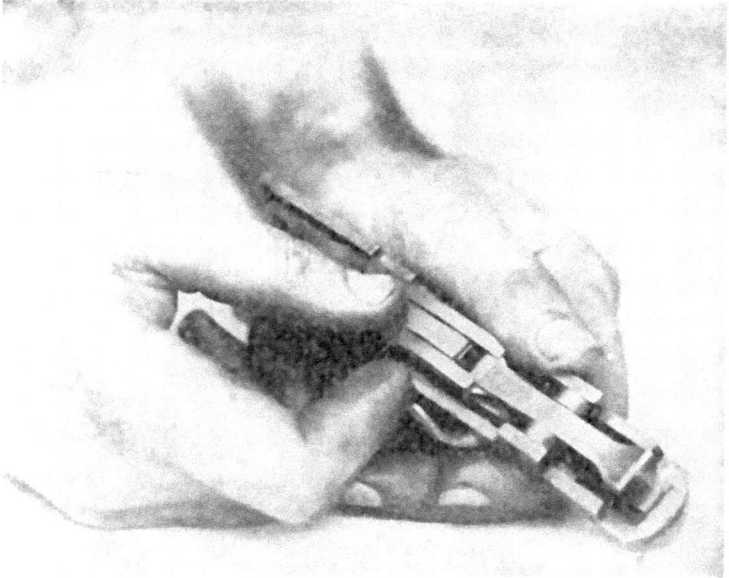

FIGURE 8.—Removing sear spring.

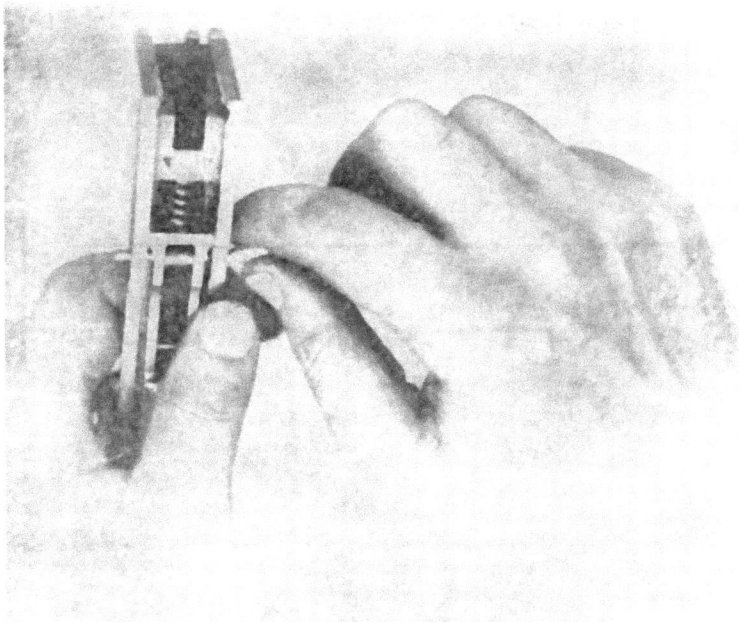

FIGURE 9.—Removing trigger pin.

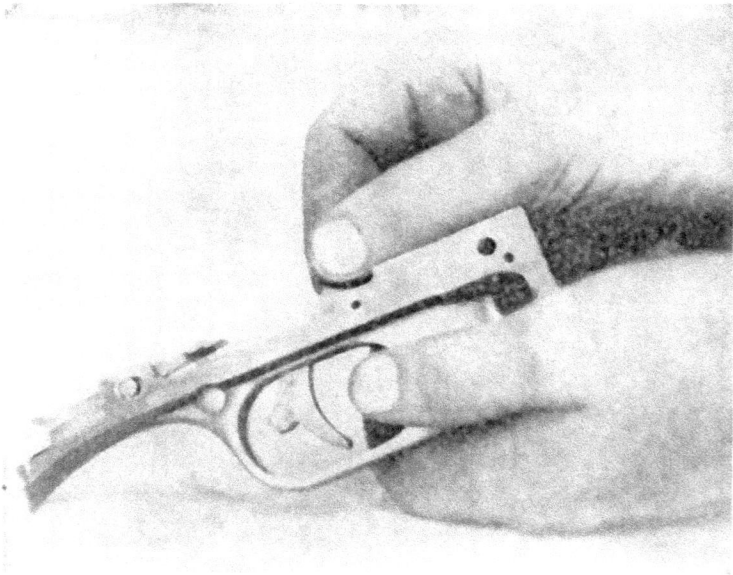

FIGURE 10.—Removing trigger and connector.

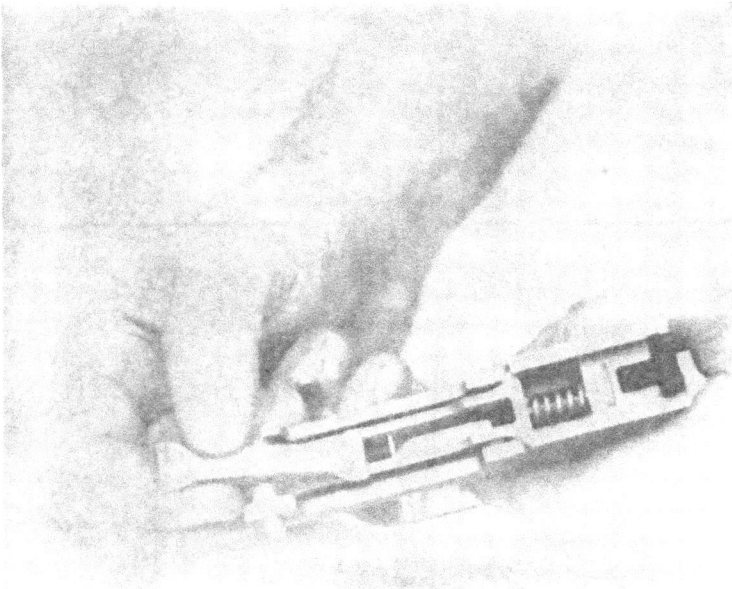

FIGURE 11.—Removing sear.

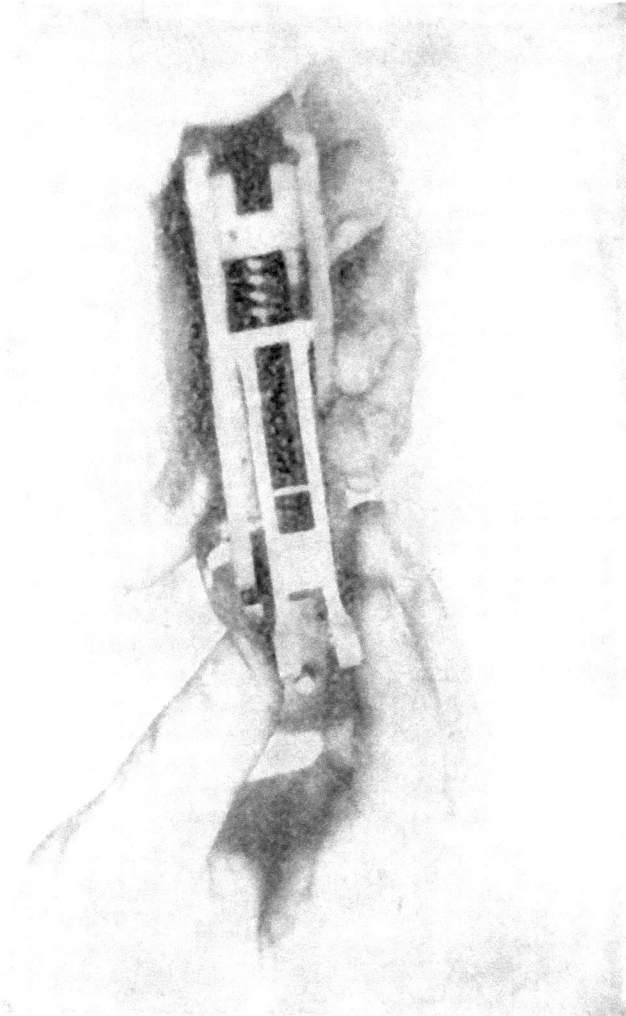

FIGURE 12.—Removing sear carrier.

(3) The forearm group, barrel group, buttstock and buffer group, sight group, and receiver group are not disassembled by using organizations. Illustrations of the buttstock and buffer group and of the sight group are shown for purposes of information only.

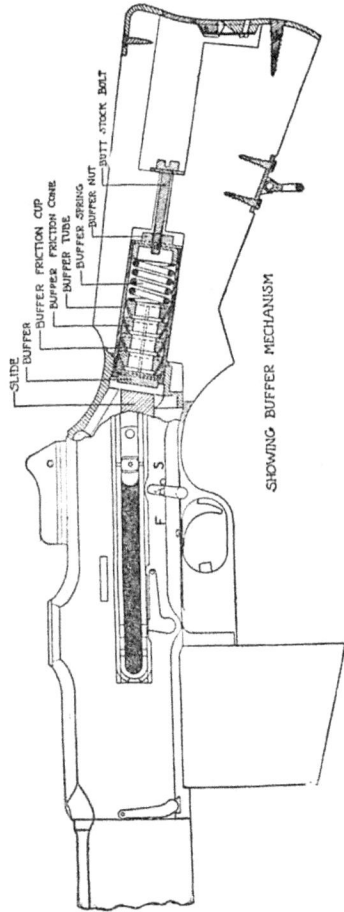

FIGURE 13.—Buttstock and buffer group.

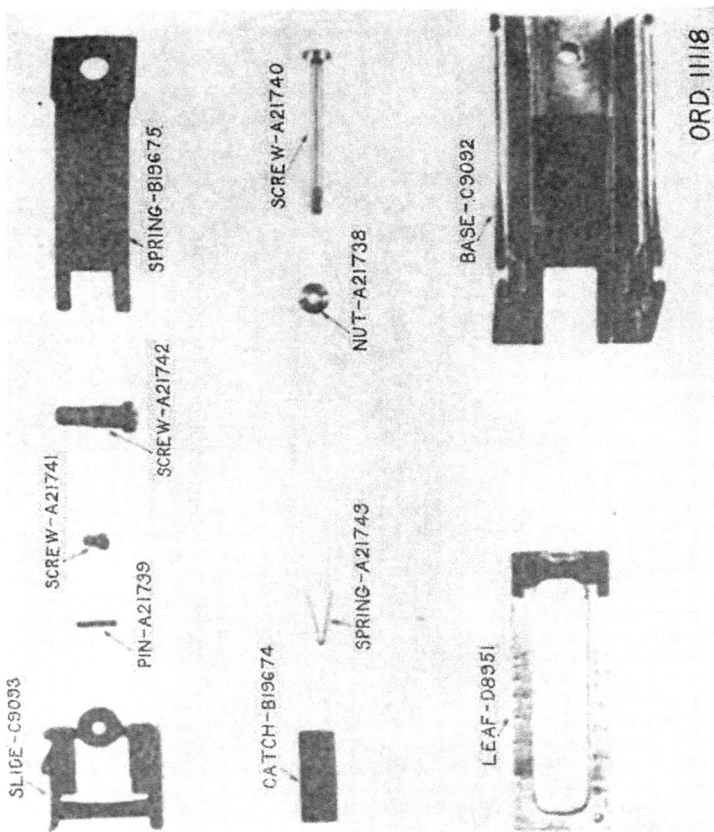

FIGURE 14.—Sight group.

■ 9. ASSEMBLING RIFLE.—The rifle and its component groups are assembled in the reverse order of their disassembly as given in paragraph 8*b*.   The following detailed explanation of the method of assembling the rifle is furnished as an aid to instructors.

a. Trigger mechanism.—Replace the change lever. Insert the ears of the change lever spring in the slots in the trigger guard and push the spring forward into place. Replace the counterrecoil spring on the counterrecoil spring guide (front of sear carrier). Insert the counterrecoil spring guide into its seat. Using the recoil spring guide (fig. 2), as a lever in the sear pin hole, pry the sear carrier forward until its rear end is held by the ears of the change lever spring. Replace the sear and force the recoil spring guide (fig. 2) through so as to register the holes in the sear, sear carrier, and trigger guard for the sear pin. By a slight pressure on the recoil spring guide push the sear carrier forward against the counterrecoil spring, thus permitting the sear pin to be easily seated in the sear pin hole. The sear pin must always be replaced before the trigger pin in order that the tension of the counterrecoil spring will always be on the sear pin. Replace the trigger and trigger pin.

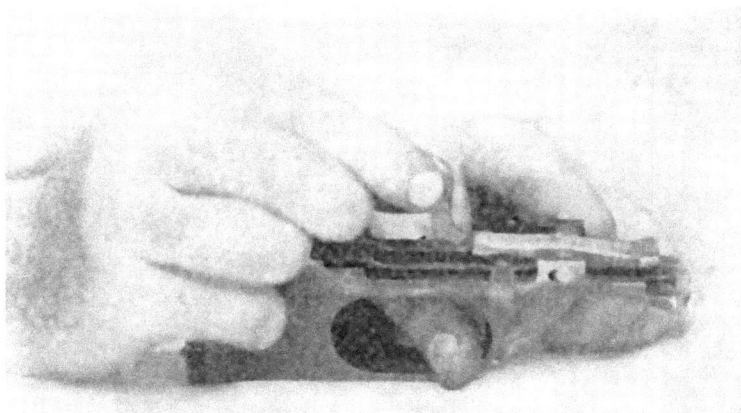

FIGURE 15.—Assembling connector.

23

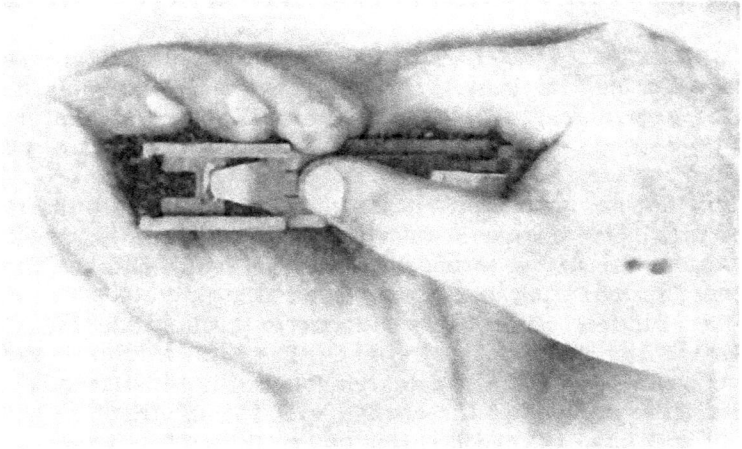

FIGURE 16.—Assembling sear spring.

FIGURE 17.—Assembling magazine catch spring.

24

Holding the connector, so that its head is in rear of the connector stop (see fig. 15) and the toe is down and to the rear, drop it into its place in the trigger. Engage the sides of the sear spring in the recesses, and press down and forward on the sear spring with the thumb of the right hand, until the front end of the spring rests in the depression stop. (See fig. 16.) Take care to see that the outside prongs of the sear spring rest on their seat on the sear, and that the middle prong rides freely in the slot formed by the walls of the sear carrier. If the middle prong rests on one of the walls, instead of riding freely between them, the trigger mechanism will not function when the barrel is inclined below the horizontal.

Replace the magazine release, magazine catch, and magazine catch pin. Replace the magazine catch spring. Insert the ejector into the recess and move it down until it is flush with the magazine catch spring. (See fig. 17.) Compress the magazine catch spring in its seat and move the ejector down until it is fully home and the ejector lock is in its position.

After the trigger mechanism has been assembled, turn the change lever to the forward position and pull the trigger. If the connector will not rise, it is not in place correctly. It should rise and snap out from under the sear. If the connector will rise but does not raise the sear, the sear spring is weak and should be replaced.

b. Operating group.—Replace the extractor spring (fig. 2). Replace the extractor into its seat in the bolt. Replace the bolt link and bolt link pin with the shoulder of the link against the flat surface of the bolt lock. Lift the bolt lock and replace the firing pin. Lay the rifle barrel down and pointing to the left so that the rifle is resting on the barrel and rear sight. With the bolt mechanism held in a perpendicular position, insert it in the receiver, forcing the end of the bolt under the ends of the bolt supports, and then press the bolt mechanism down so as to lie flat in its place. Push the bolt mechanism forward, swing the bolt link down, then replace the slide and push it all the way back. With the hammer resting between the thumb and the forefinger, lower and seat it properly in its seat in the slide, push the bolt lock fully into its locking recess and push the slide forward. With the thumb and forefinger of the right hand, aline the hammer pin holes in the bolt link, hammer, and slide with the hammer pin hole in the side of the receiver. The recoil spring guide will be found a convenient aid in the operation. Insert the hammer pin to the right until only one-fourth of an inch of the hammer pin protrudes from the receiver. Replace the operating handle ((88), fig. 1 ②). Tap the end of the protruding hammer pin with sufficient force to drive it home. Replace the recoil spring (fig. 2) and guide. With

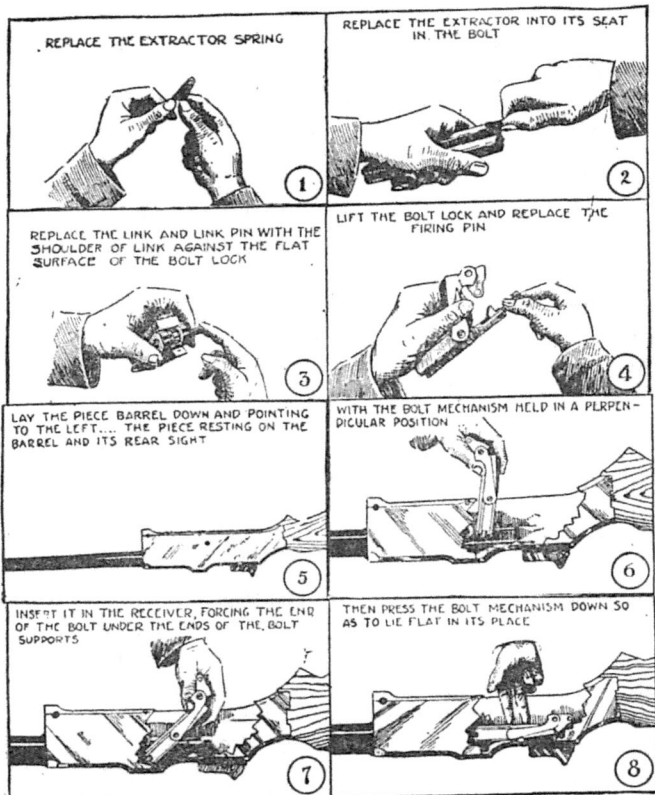

FIGURE 18.—Method of assembling.

the end of the index finger on the checkered end of the recoil spring guide head, turn it until it is properly seated. Holding the right thumb against the forward end of the receiver will facilitate this operation. Replace the trigger guard group and trigger guard retaining pin. Cock the piece. Slide the gas cylinder tube and forearm ((118) fig. 1 ②) to the rear of the gas piston. Replace the gas cylinder tube retaining pin. Test the piece by pulling the trigger.

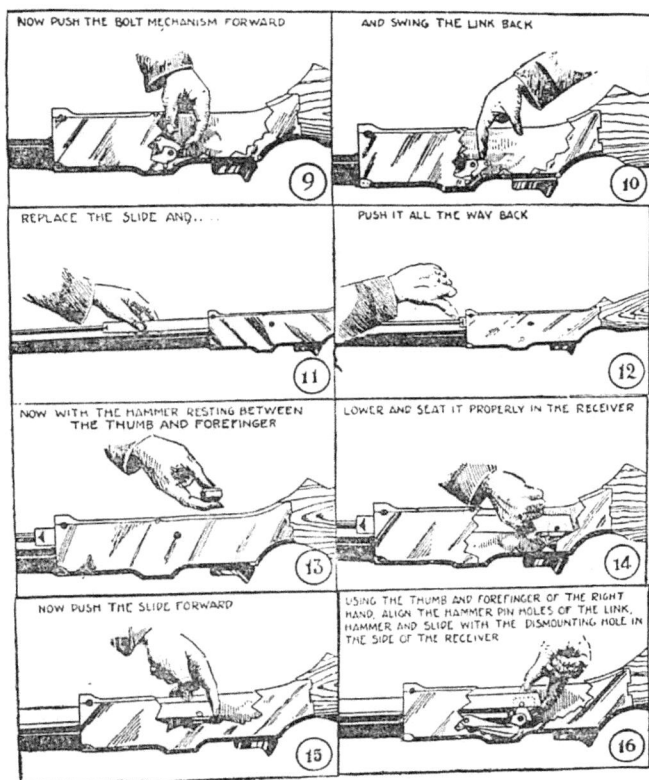

FIGURE 18.—Method of assembling—Continued.

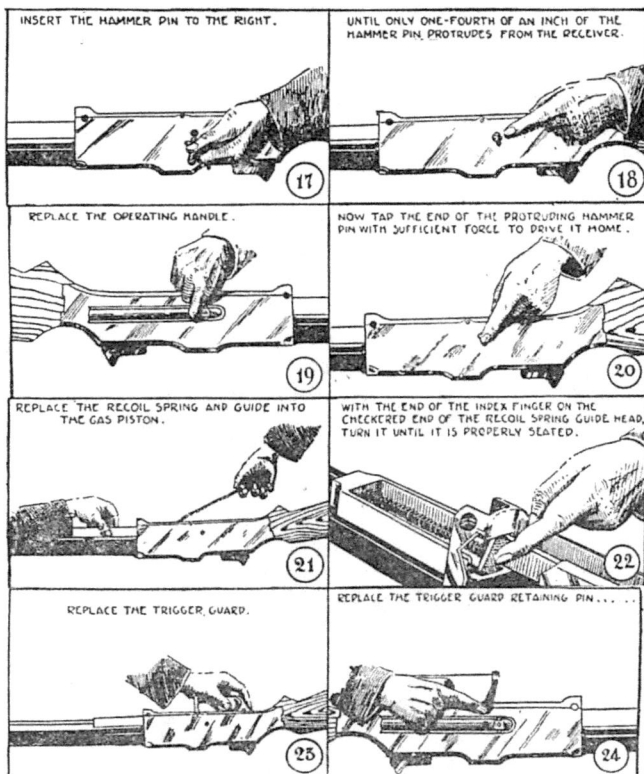

FIGURE 18.—Method of assembling—Continued.

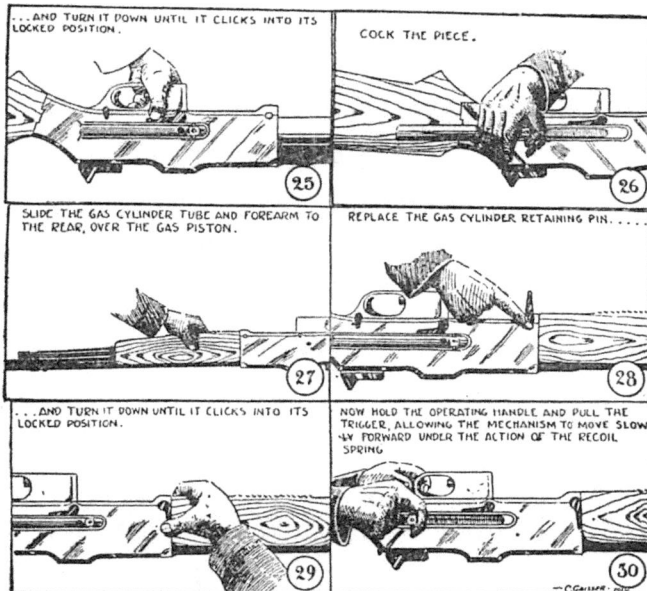

FIGURE 18.—Method of assembling—Continued.

■ **10. To Remove Firing Pin Without Disassembling.**—To remove the firing pin, lay the rifle on the table, barrel down, muzzle to the front. Remove the trigger mechanism. Place rim of cartridge under bolt guide (fig. 19 ①). Pull operating handle to rear and hold mechanism back. Steady the cartridge with the thumb and forefinger of the right hand (fig. 19 ②). It may be necessary to exert a slight downward pressure on the nose of the cartridge in order to pull the bolt guide out far enough to free the bolt. Push down on the bolt link, causing the bolt to break at the bolt lock pin (fig. 19 ③). Allow the mechanism to go forward until it stops. Change the firing pin. Pull the operating handle to rear again, and push the bolt into position (fig. 19 ④).

29

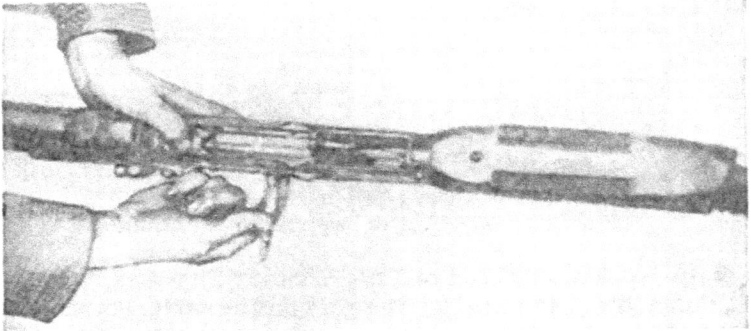

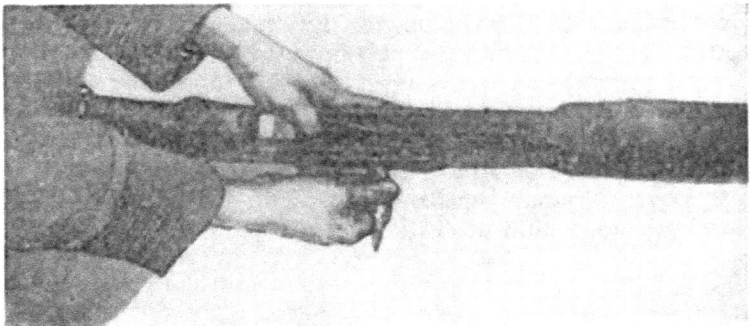

FIGURE 19.—To remove firing pin without disassembling rifle.

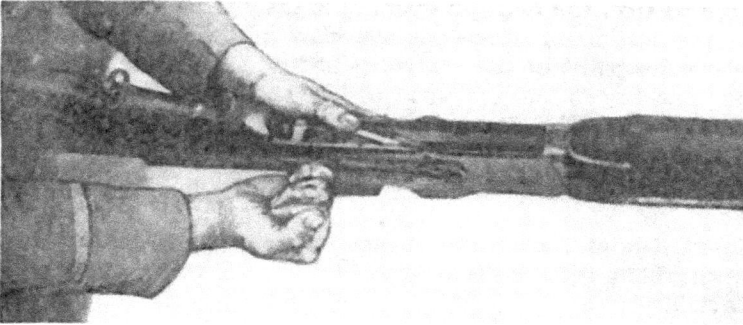

FIGURE 19.—To remove firing pin without disassembling rifle—
Continued.

■ 11. To REMOVE AND REPLACE EXTRACTOR WITHOUT DISASSEM-
BLING.—*a. Removal.*—Draw the mechanism to the rear and
insert an empty cartridge case between the bolt and chamber, exposing the extractor (fig. 20 ①). Lay the rifle on its

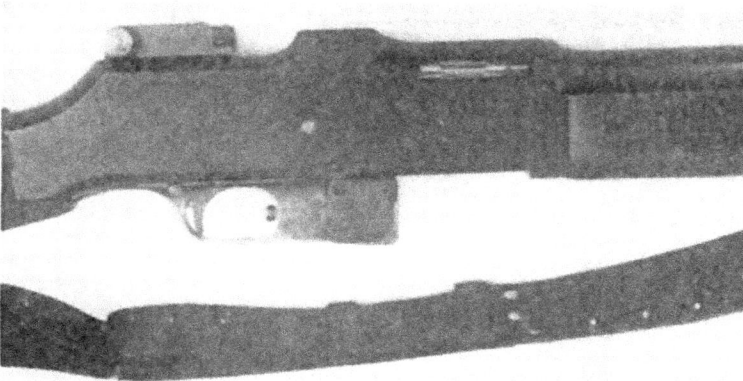

FIGURE 20.—To remove and replace extractor without disassembling
rifle.

31

side so that the ejection opening is up. With the forefinger of the left hand, force out the claw of the extractor, then place the point of the cartridge behind the extractor shoulder and pry it forward until the extractor is free of the recess (fig. 20 ②). Remove the extractor spring.

  b. Replacement.—Insert the short end of the extractor spring in the hole in the shank of the extractor so that the long end of the spring is along the slot in the extractor. Insert the extractor and spring in the end of the bolt and push them into position (fig. 20 ③). Remove the empty cartridge case.

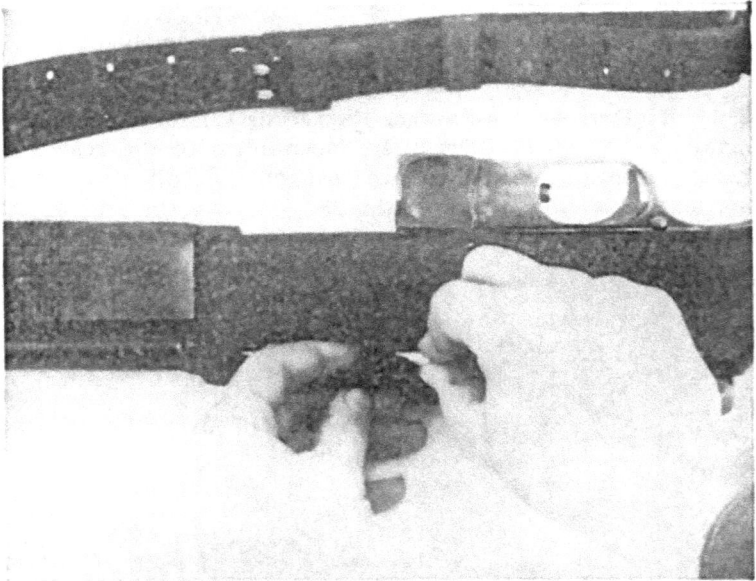

②

FIGURE 20.—To remove and replace extractor without disassembling rifle—Continued.

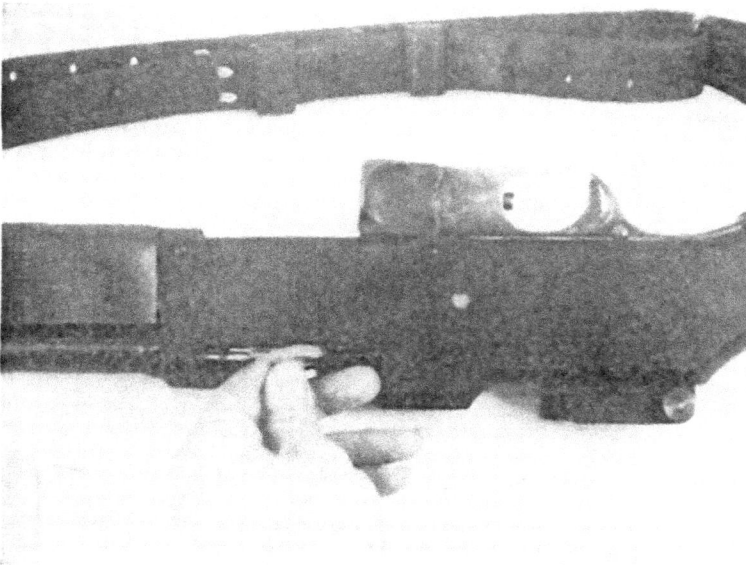

FIGURE 20.—To remove and replace extractor without disassembling
rifle—Continued.

■ 12. DISASSEMBLING AND ASSEMBLING MAGAZINE.—Raise the
rear end of the magazine base until the indentations on it
are clear, then slide it to the rear. The magazine follower
and spring will then fall out. It is assembled in reverse
order. (See figs. 21 to 24, incl.)

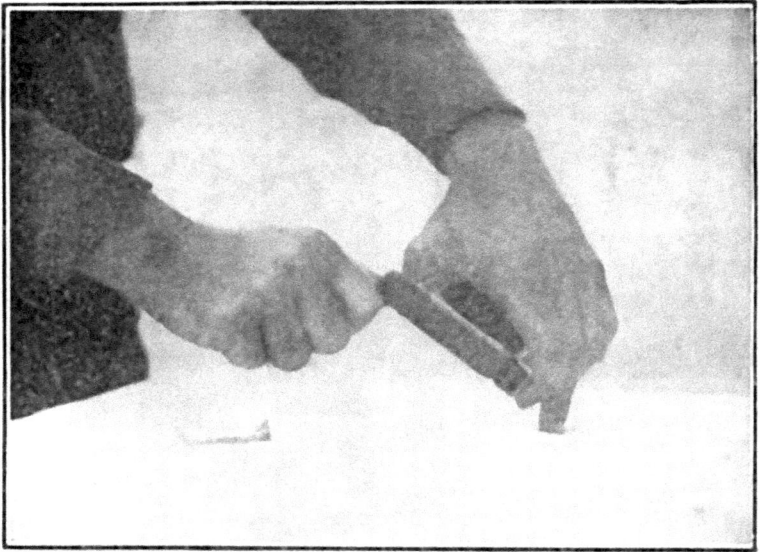

FIGURE 21.—Disassembling magazine, first operation.

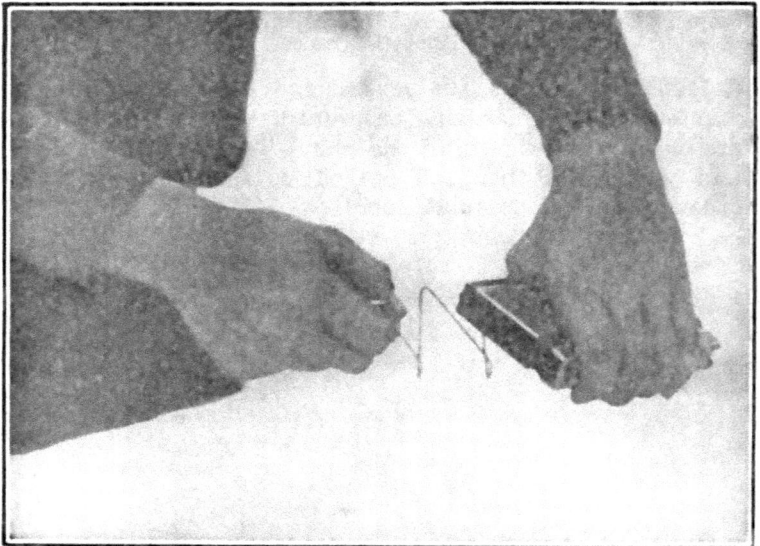

FIGURE 22.—Disassembling magazine, second operation.

FIGURE 23.—Assembling magazine, first operation.

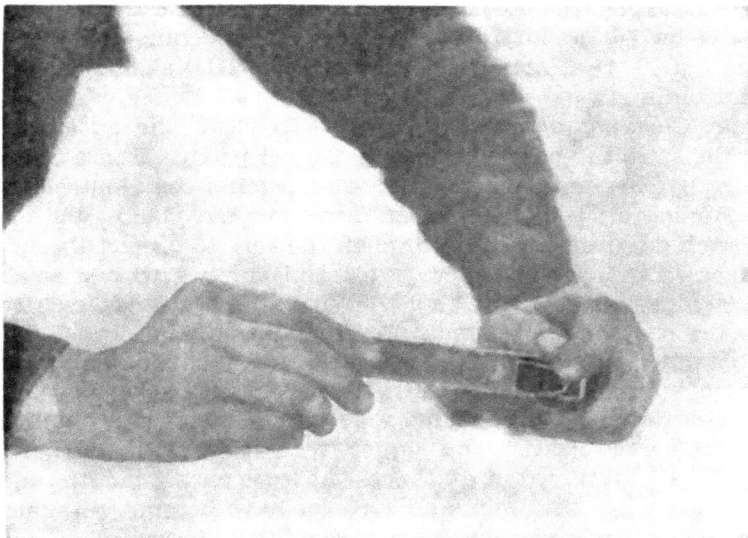

FIGURE 24.—Assembling magazine, second operation.

SECTION III

CARE AND CLEANING OF THE RIFLE

■ 13. GENERAL.—*a. Scope.*—(1) Care and cleaning includes the care of the automatic rifle necessary to preserve its condition and appearance under all conditions and at all times. Automatic rifles in the hands of troops should be inspected daily to insure proper condition and cleanliness.

(2) Automatic rifles should be disassembled only to the extent necessary for cleaning and proper lubrication.

b. Lubrication and lubricants.—(1) Proper oiling is second in importance only to intelligent cleaning. It is a vital necessity for the working parts, but the oil should be used sparingly; wiping with a well-oiled rag is the best method. Oil all bearing surfaces of the rifle before firing.

(2) If the rifle is to be fired in areas where the temperature is 45° F., or above, sperm oil (U. S. Army Specification No. 2–45A) should be used for oiling, when available. When not available, motor oil, weight 20, or any light-grade machine oil, may be used in an emergency.

(3) If the rifle is to be fired in areas where the temperature is below 45° F., aircraft instrument and machine-gun lubricating oil (U. S. Army Specification No. 2–27D) should be used for oiling.

c. Cleaning of automatic rifle.—To clean the automatic rifle, swab the bore with an oily flannel patch. Repeat with dry patches until several successive patches come out clean. (For cleaning the bore after firing, see par. 14c.) Push a patch dampened with oil through the bore to protect its surface. Dust out all screw heads and crevices with a small cleaning brush or small stick. Wipe all metal surfaces with a dry cloth to remove moisture, perspiration, and dirt. Wipe the outer surfaces of the automatic rifle, including the forearm, with a lightly oiled cloth, then clean with a soft dry one. Immediately after cleaning, wipe all the metal parts with a lightly oiled cloth. This protective film on all metal parts will be maintained at all times. At least once a month, and always after the stock and forearm have become wet, they should be rubbed thoroughly with a little linseed oil in the palm of the hand. Rub oil in until dry. Use only castile soap or saddle soap for cleaning or softening the sling.

■ 14. ADDITIONAL RULES FOR CARE OF AUTOMATIC RIFLE PRE-PARATORY TO, DURING, AND AFTER FIRING.—*a. Preparatory to firing.*—(1) Remove the protective film of oil from bore and chamber.

(2) Work slide back and forth several times to see that it moves freely.

(3) Verify proper setting of gas port.

(4) Test trigger mechanism at Safe (S), Semiautomatic (F), and Automatic (A).

(5) Examine magazines. It is imperative that magazines be given the best of care and kept in perfect condition. They should be disassembled, wiped clean and dry, and thinly coated with oil. Much dirt gets into them through careless handling on the ground during range or other firing. They must be kept free from dirt and rust, which hinder their operation by making the spring and follower stick. Care must be exercised in the handling of magazines to avoid dent-ing or bending them. The greatest possible care should be taken to prevent any damage to the lips of the magazine or to the notch for the magazine catch.

b. During firing.—(1) Keep bore free from dust, dirt, mud, or snow.

(2) Keep chamber free from oil or dirt.

(3) Keep moving parts oiled.

(4) Clean bore and gas system frequently while still hot. The neglect of this precaution is a frequent cause of stop-pages.

(5) Clean chamber frequently with chamber brush by in-serting the cleaning brush through the ejection opening in the receiver.

c. After firing.—(1) The bore of the rifle will be thoroughly cleaned by the evening of the day on which it is fired, and similarly cleaned for the next 3 days.

(2) The bore is cleaned after firing by swabbing it with a flannel cleaning patch saturated with hot water and sal soda or issue-soap solution. Repeat with several patches. Plain water, hot or cold, should be used when both soda and soap

are lacking.   While the bore is still wet, run the metal brush through it several times.   Follow this with dry patches until several patches come out clean and dry, then push a patch saturated with oil through the bore to protect its surface.

(3) Clean the chamber with the chamber-cleaning brush, wipe clean with a cloth, and oil lightly.

(4) Clean the gas system by first disassembling the rifle. Remove the gas cylinder.   Insert the smooth end of the body of the gas-cylinder tool into the gas cylinder.   As it is advanced toward the cylinder head turn it to the right.   As it reaches the head, apply additional pressure to the tool and give it a few turns to cut the carbon from the inside surface of the piston head.   Withdraw and reverse the tool.   Using the recess cutter as a gage, remove the carbon from the recesses at the forward end of the interior of the gas cylinder.   With the drift point, clean the gas ports in the barrel, gas-cylinder tube, and gas cylinder.   Scrape the carbon from the face of the piston with the front cutting edge of the tool body and remove the deposit from between the piston rings with the drift point.   Wash with hot water and soap or sal-soda solution (if not available, use plain water), dry thoroughly, and oil lightly.

(5) Clean magazines by disassembling, wiping, oiling, and reassembling.

■ 15. STORAGE.—*a. Preparation of automatic rifles for long-term storage.*—Automatic rifles should be cleaned and prepared with particular care.   The bore, all parts of the mechanism, and the exterior of the rifles should be thoroughly cleaned and then perfectly dried with rags.   In damp climates particular care must be taken to see that the rags are dry.   After drying a part, the bare hands should not touch that part.   Special care should be taken to insure that the gas system is thoroughly cleaned and that the gas ports are free from fouling.   All metal parts should then be heavily coated with rust-preventive compound.   Then handling the rifle by the stock and forearm only, it should be placed in the packing chest, the wooden supports at the butt and muzzle having previously been painted with rust-preventive compound.   A rifle contained in a cloth or other cover, or with a plug in the bore, will not be placed in storage.

Such articles collect moisture which causes the weapon to rust.

b. Cleaning of automatic rifles as received from long-term storage.—Automatic rifles received from storage are completely coated with rust-preventive compound. Use dry-cleaning solvent to remove all traces of this compound, particular care being taken that the gas system, gas ports, firing pin, and all recesses in which springs or plungers operate are cleaned thoroughly. After using the dry-cleaning solvent, make sure it is completely removed from all parts by wiping with light-colored cloths until no staining of the cloth occurs. The bore and chamber of the barrel must be thoroughly cleaned. All surfaces having been thoroughly cleaned, they should then be protected with a thin film of lubricating oil applied with a rag.

NOTE.—Failure to clean the gas system, the firing pin, and the recess in the bolt in which it operates may result in gun failure at normal temperatures and will most certainly result in serious malfunctions if the rifles are operated in low-temperature areas, as rust-preventive compound and other foreign matter will cause the lubricating oil to congeal on the mechanism.

SECTION IV

FUNCTIONING

■ 16. OBJECT.—This section is designed to provide a nontechnical description of the functioning of the rifle. The object of instruction in this subject is to lead the soldier to an understanding of the simple functioning of his weapon without emphasis on memorizing the matter of the text.

■ 17. WHEN TAKEN UP.—Instruction in functioning will be taken up after instruction in the disassembly, assembly, care, and cleaning of the rifle.

■ 18. USE OF DUMMY CARTRIDGES.—The corrugated type of dummy cartridge (cartridge, dummy, caliber .30, M1906–corrugated) may be used for instruction in functioning. The use of the slotted type of dummy cartridge (cartridge, dummy, range, caliber .30, M1) is prohibited. Special care must be exercised in the use of dummy cartridges that they do not introduce dirt or grit into the chamber of the rifle.

39

■ 19. EXPLANATION.—All automatic weapons have mechanical means for performing the following functions: extraction, ejection, feeding, locking the breech while there is a high pressure in the bore, and igniting the cartridge.

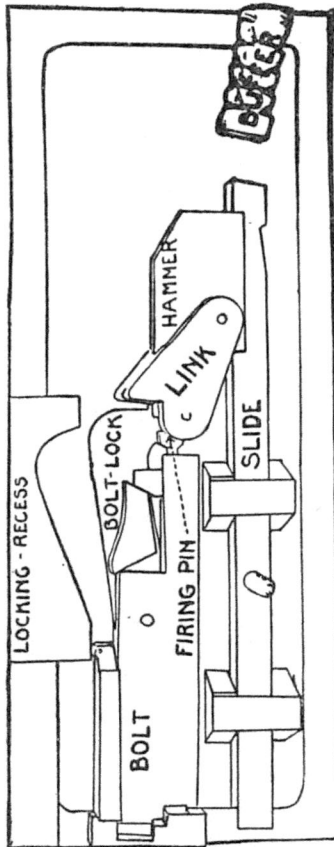

FIGURE 25.—Bolt action.

Operations such as extraction and ejection are performed by various cams, lugs, and springs, and the energy necessary to perform this work and to overcome friction in the rifle is derived from the explosion of the powder in the chamber.

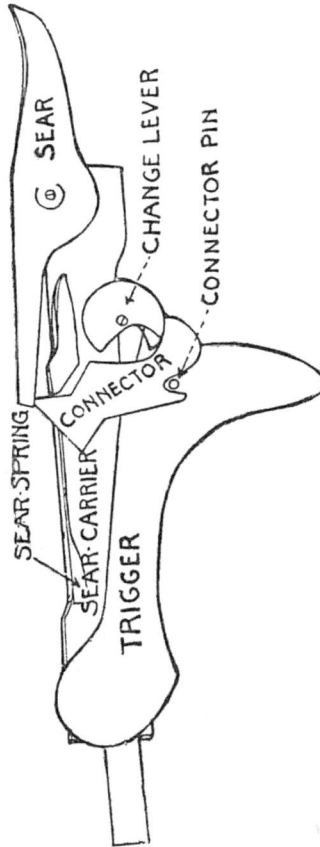

FIGURE 26.—Trigger mechanism.

■ 20. CYCLE.—*a.* The functioning of the automatic rifle is divided into two phases based on the operation of the mechanism when a shot is fired. These two phases are the rearward movement and the forward movement. The ignition of the cartridge in the chamber marks the division of the two phases.

b. The operations which take place in the rearward movement are—

(1) Action of gas.
(2) Movement of slide to rear.
(3) Unlocking.
(4) Withdrawal of firing pin.
(5) Extraction.
(6) Ejection.
(7) Termination of first phase.

c. The operations which take place in the forward movement are—

(1) Action of recoil spring.
(2) Feeding.
(3) Locking.
(4) Ignition.
(5) Termination of second phase.

■ 21. DESCRIPTION OF CYCLE.—*a. Rearward movement.*—(1) *Action of gas.*—A cartridge having been ignited, the bullet under the pressure of the expanding powder gases travels through the barrel, and when it reaches a point 6 inches from the muzzle, it passes a port in the bottom of the barrel. The barrel pressure, which at this instant is still very high, seeks this first vent. Alined with the barrel port are other similar ports in the gas cylinder tube bracket, gas cylinder tube, and gas cylinder. The port in the gas cylinder is the smallest and serves to throttle the barrel pressure. The ports in the gas cylinder lead radially into a well about $\frac{1}{8}$ inch in diameter in the head of the gas cylinder. The throttled barrel pressure is conducted through this well to the gas piston plug. This pressure acts on the piston for the very short time which it takes for the bullet to travel the 6 inches distance from the barrel port to the muzzle. Its effect is that of a sudden severe blow on the piston plug. Under the impact of this blow the gas piston is driven to the rear, carry-

ing the slide with it. When the piston has traveled about
$9/16$ inch backward, the bearing rings on its head and the gas
piston plug pass out of the cylinder.   The gas expands around
the piston head into the gas cylinder tube, and is exhausted

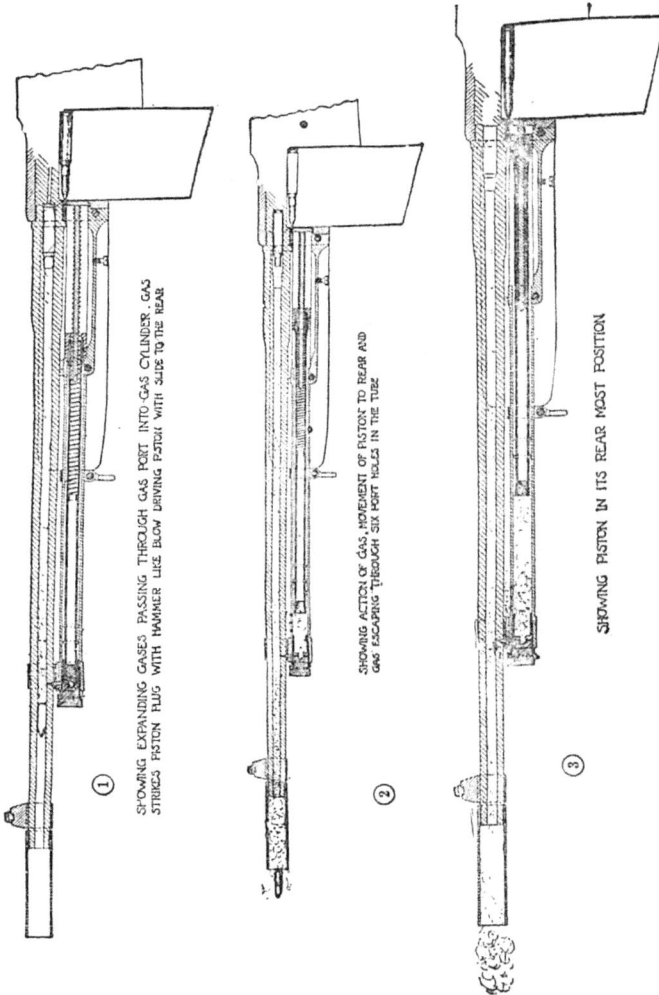

FIGURE 27.—Action of gas on piston.

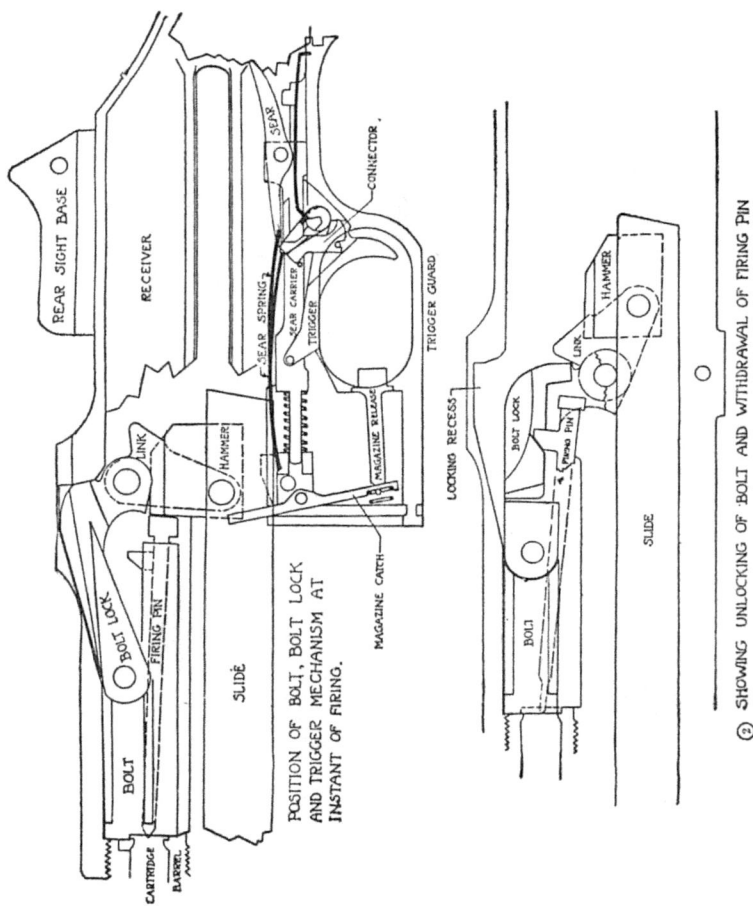

REAR SIGHT BASE

RECEIVER

SEAR

CONNECTOR

SEAR SPRING

REAR CARRIER

TRIGGER

TRIGGER GUARD

LINK

HAMMER

MAGAZINE RELEASE

MAGAZINE CATCH

BOLT LOCK

FIRING PIN

SLIDE

BOLT

CARTRIDGE

BARREL

POSITION OF BOLT, BOLT LOCK
AND TRIGGER MECHANISM AT
INSTANT OF FIRING.

LOCKING RECESS

HAMMER

LINK

BOLT LOCK

FIRING PIN

SLIDE

BOLT

② SHOWING UNLOCKING OF BOLT AND WITHDRAWAL OF FIRING PIN

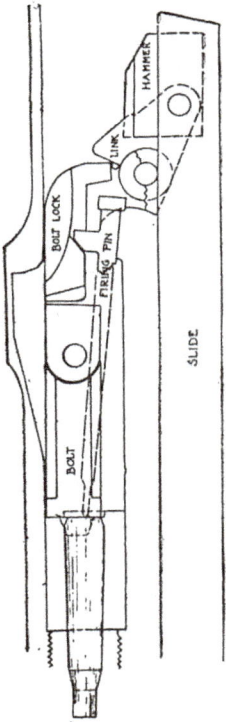

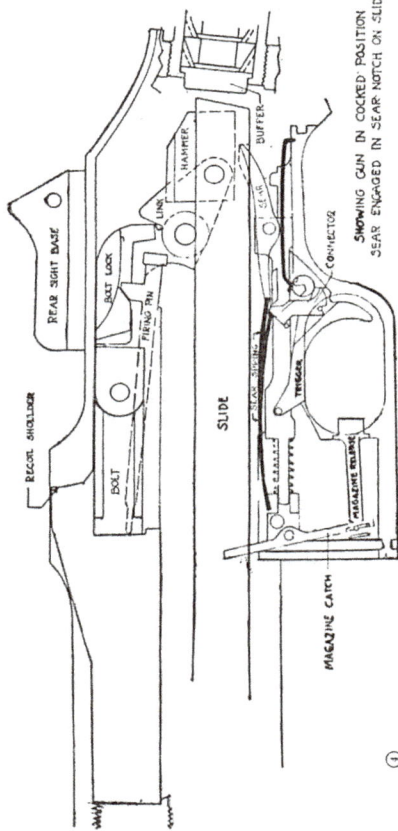

FIGURE 28.—Details of functioning.

through the six portholes in the tube. The gas is prevented from traveling back through the gas cylinder tube by the two rings on the piston, about $\frac{5}{8}$ inch apart and $1\frac{1}{4}$ inches from the piston head. These rings also serve as bearings to hold the front end of the piston in the center of the gas cylinder tube after the piston head has passed out of the gas cylinder.

(2) *Slide.*—As the piston is forced back it carries the slide with it. The first and immediate result of the backward movement of the slide is to begin the compression of the recoil spring, thereby storing energy for the forward action.

(3) *Unlocking.*—The hammer pin is slightly in advance of the bolt-link pin, about 0.19 inch. The center rib of the hammer is very slightly in rear of the head of the firing pin. When the slide begins its motion to the rear, it imparts no motion whatever to the bolt and bolt lock. The slide moves back 0.19 inch, and its only effect during this travel is to carry the hammer from the firing pin and the hammer pin directly under the bolt-link pin. At this point the unlocking begins, the bolt link revolves forward about the hammer pin, drawing the bolt lock down and to the rear. The motion of the lock and bolt, which is zero at the instant the hammer pin passes under the bolt-link pin, accelerates from this point until the slide has traveled about 1.2 inches, at which latter point the bolt lock is drawn completely down out of the locking recess and away from the locking shoulder of the receiver. It is now supported in front of the bolt supports. The front upper shoulder of the bolt link has revolved forward and bears upon the locking shoulder of the bolt lock. These two influences prevent the bolt lock from revolving down below the line of backward travel of the bolt.

(4) *Withdrawal of firing pin.*—As the bolt lock revolves down from its locked position a cam surface in the slot in the rear bottom side of the bolt lock comes in contact with a similar cam surface on the firing-pin lug. This action cams the firing pin from the face of the bolt.

(5) *Extraction.*—The backward motion of the bolt begins when the bolt lock has been drawn down so that the circular cam surface on its underside is operating on the rear shoulders of the bolt supports. This produces a strong lever action which slowly loosens the cartridge case. The backward travel of the bolt has been slight, only about 5/32 inch when the firing pin is withdrawn; its travel is about 11/32 inch when the bolt lock is completely drawn down. From this point the bolt moves to the rear, drawn by the bolt lock and bolt link, with the same speed as the slide and carries with it the empty cartridge case, which is held firmly in its seat on the face of the bolt by the extractor. The extractor is on the upper right-hand side of the bolt next to the ejection opening in the receiver. A slot cut in the left side of the bolt lock near the back end passes over the bolt guide, which supports the bolt lock and bolt when they are in the rear position.

(6) *Ejection.*—When the slide reaches a point about ¼ inch from the end of its travel, the base of the cartridge case strikes the ejector. This action causes the cartridge case to be pivoted with considerable force about the extractor, and through the ejection opening in the receiver. The front end of the cartridge case passes first out of the receiver and is pivoted so that it strikes the outside of the receiver at a point about 1 inch in rear of the ejection opening. It rebounds from the receiver toward the right front.

(7) *Termination of rearward movement.*—The rearward motion is terminated when the rear end of the slide strikes the buffer. The slide moves forward $\frac{1}{10}$ inch, after striking the buffer, under the action of the recoil spring, but if the sear nose is not depressed, it engages the sear notch on the slide, and the piece is cocked for the next shot.

NOTE.—The motion of the bolt, bolt lock, and bolt link mechanism begins slowly at first and does not attain the speed of the slide until the latter has traveled about 1¼ inches backward. This is a very important characteristic of the rifle since on this account the mechanism is not subjected to an excess strain due to a sudden start at the instant the gas impinges upon the piston. This slow start delays the opening of the chamber sufficiently to allow the high barrel pressure to decrease.

b. Forward movement.—(1) *Action of recoil spring.*—The sear nose is depressed, disengaging the sear, and the slide moves forward under the action of the recoil spring. The position of the bolt link pin is slightly below a line joining the bolt lock pin and the hammer pin; therefore, as the slide starts forward the joint at the bolt link pin has a tendency to buckle downward. It is prevented from doing this by the tail of the feed rib on the bolt which extends backward under the bolt lock, also by the contact of the upper front shoulder of the bolt link with the locking surface of the bolt lock. Since the joint cannot buckle, the entire mechanism moves forward with the slide. When it has traveled about ¼ inch, the front end of the feed rib impinges on the base of the top cartridge, which the magazine spring and lips are holding up in its path.

(2) *Feeding.*—The cartridge is carried forward about ¼ inch when the nose of the bullet strikes the bullet ramp or guide on the breech of barrel and is deflected upward toward the chamber. This action also guides the front end of the cartridge from under the magazine lips. The base of the cartridge approaches the center of the magazine, where the lips are cut away and the opening enlarged, and at this point is forced out of the magazine by the magazine spring. The base of the cartridge slides across the face of the bolt and under the extractor. Should the cartridge fail to slide under the extractor, the extractor will snap over its head as the bolt reaches the forward position. When the cartridge is released by the magazine, the nose of the bullet is so far in the chamber that it is guided by the chamber from this point on.

(3) *Locking.*—When the slide is about 2 inches from its forward position, the circular cam surface on the under side of the bolt lock begins to ride over the rear shoulders of the bolt supports, and the rear end of the bolt lock is cammed upward. The bolt link pin passes up above the line joining the bolt lock pin and hammer pin. The joint at the bolt link pin now has a tendency to buckle upward, and the bolt lock· be-

ing opposite the locking recess in the receiver, is free to, and does, pivot upward about the bolt lock pin. The bolt link revolves upward about the hammer pin, forcing the bolt lock up, and a rounded surface on the bolt lock just above the locking face slips over the locking shoulder in the receiver, giving the lock a lever action which forces the bolt home to its final position. The two locking surfaces on the bolt lock and the receiver register as the hammer pin passes under the bolt link pin.

(4) *Igniting cartridge.*—The lug on the firing pin is buried in the slot on the underside of the bolt lock at all times except when the bolt is locked in the forward position. Therefore, the firing pin is locked away from the face of the bolt during all the rearward and forward motion of the bolt. When the hammer pin passes under the bolt link pin, the firing pin has been released by the bolt lock. The slide and hammer move forward about $\frac{1}{10}$ inch farther, and the center rib of the hammer strikes the head of the firing pin, driving it forward and igniting the cartridge.

(5) *Termination of second phase.*—The forward end of the slide strikes a shoulder at the rear end of the gas cylinder tube which terminates the forward motion. The forward motion is not terminated by the hammer on the firing pin. This can be seen by examining the head of the firing pin when the gas cylinder tube is assembled to the receiver, and the bolt mechanism is in the forward position. The firing pin has still about $\frac{1}{16}$-inch clearance from its extreme forward position.

NOTE.—The locking shoulder of the receiver is inclined forward. Its surface is perpendicular to the line through the bolt lock which the shock of the explosion follows; therefore, the force of this shock is exerted squarely against the normal surface. The speed of the bolt mechanism is slowed down gradually from the instant that the bolt lock starts to rise until the hammer pin passes under the bolt link pin, when its speed is zero.

■ 22. FUNCTIONING OF BUFFER.—*a.* The buffer system consists of a tube, in which are placed successively, from front to rear, the buffer head, a brass friction cup with concave interior which is split to allow for expansion, and a steel cone to fit into the cup. Four of these cups and cones are placed one after the other in series. In rear of these is the buffer spring, and finally the buffer nut, which is screwed into the end of the tube and forms a seat for the spring.

b. Action.—The buffer head, struck by the rear end of the slide, moves to the rear, forcing the cups over the cones and causing them to expand tightly against the tube, consequently producing considerable friction as the cups move back and compress the buffer spring. Thus the rearward action of slide is checked gradually and there is practically no rebound. The spring returns the buffer head and friction cups and cones to their original positions.

■ 23. FUNCTIONING OF TRIGGER MECHANISM.—*a.* The trigger mechanism has three settings:

(1) *Automatic (A).*—When so set, the sear is depressed as long as the trigger is held back and the piece will continue firing until the magazine is emptied.

(2) *Semiautomatic (F).*—When so set, the sear is depressed, thereby disengaging the sear and sear notch when the trigger is pulled, but the mechanism is so constructed that the sear rises and engages in the sear notch when the slide comes back again, and the sear and sear notch will not disengage until the trigger is fully released and then pulled. With this setting the piece fires one shot for each pull and release of the trigger.

(3) *Safe (S).*—When so set, the sear cannot be released from the sear notch by pulling the trigger.

b. The action of the trigger mechanism is taken up in phases, and should be followed on the mechanism itself as the explanation proceeds. Have the trigger guard disassembled completely. Study the shape of the change lever and note the following:

(1) It is a bar about ¼ inch in diameter.

(2) It has three shallow longitudinal slots cut on top of the bar as the handle is held vertically.

(3) The side of the bar is slotted in such a way as to leave a little tongue of metal in the center and at the lower edge of the slot.

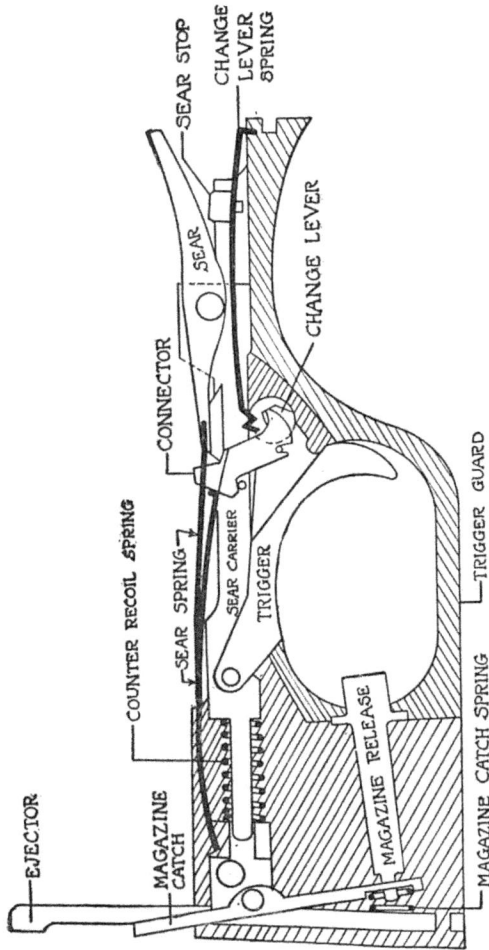

① Showing trigger mechanism single-shot firing. The connector cammed forward from under the forward end of the sear by the under-cam surface of the sear carrier permits the forward end of the sear to return to position under tension of sear spring, causing the rear end of the sear to engage in the sear notch of the slide.

FIGURE 29.—Functioning of trigger mechanism.

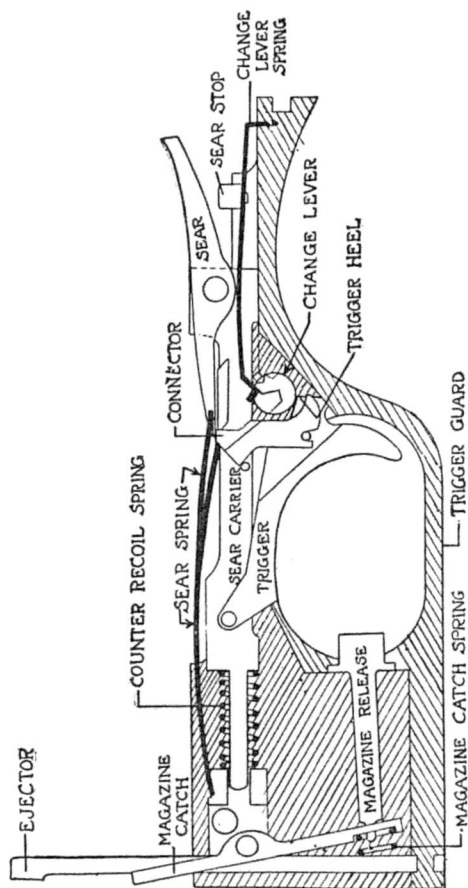

② Showing trigger mechanism set on safe. Cylindrical portion of the change lever resting over the heel of the trigger prevents the upward movement of the trigger and the releasing of the sear.

FIGURE 29.—Functioning of trigger mechanism—Continued.

■ 24. SETTING CHANGE LEVER.—*a*. Assemble the change lever and spring to the trigger guard. The toe of the change lever is seated in one of the longitudinal slots on the change lever, and as the lever is turned from one position to another it seats in the other slots. The only function of the spring and the longitudinal slots is to hold the change lever in the position in which it is set.

b. Assemble the trigger and pin to the guard.

c. Turn the change lever to rear or safe position. In this position the slot is turned slightly upward, and the full surface of the bar is on the bottom. Pull the trigger. The rear top end of the trigger is slotted longitudinally, and the metal on each side of the slot forms two shoulders which rise against the bottom of the change lever bar.

d. Push the change lever over to the vertical position, which is the automatic setting. Pull the trigger. The slot in the change lever is now turned to the front, and the two shoulders of the trigger, which before engaged the full surface of the change lever bar, now are free to pass up into the slot of the change lever; also the tongue of metal on the bottom of the change lever slot passes through the longitudinal slot in the end of the trigger.

e. Push the change lever forward to the semiautomatic position.

f. The slot is now turned partially down and when the trigger is pulled the rear end of the trigger passes up into the change lever slot; also the tongue of metal in the bottom of the change lever slot is now turned back and does not pass through the slot in the end of the trigger as it did in the automatic position.

g. Observe the shape of the connector. It is shaped like a boot with a toe and heel. It has a flat surface that slopes down and toward the front from the head. (Sear spring ramp.) In rear of the head the profile extends straight downward for about ⅛ inch, then slopes slightly to the rear for 0.12 inch. (Sear carrier ramp.) This last slope is used in a cam action to be explained later. The function of the narrow, flat top surface of the connector is to raise the forward end of the sear until cammed out from under the latter.

h. Place the connector on the connector pin and set change lever to safe. Pull the trigger. The connector is not raised, for the obvious reason that the trigger itself cannot be raised because the change lever bar is in its way.

i. Turn the change lever to the automatic position. Pull the trigger. The head of the connector is raised and held in a vertical position and cannot be tipped forward. The tongue on the change lever engages the toe of the connector as the trigger is pulled and holds the connector upright.

j. Turn the change lever to the semiautomatic position. Pull the trigger. The tongue on the change lever now does not engage the toe of the connector, and the head of the connector can now be tipped forward.

k. Note the cross pin on the sear carrier called the connector stop; also note that just in rear of the connector stop and on the underside of the sear carrier is an inclined surface sloping upward in the metal which joins the two sides of the sear carrier. This surface has a cam action with the above-mentioned surface on the connector.

l. Completely assemble the trigger mechanism.

m. Note that the center leaf of the sear spring presses on the front sloping surface of the connector and tends to press the head of the connector backward. Set the change lever on "safe" and pull the trigger. The head of the connector is not raised above the sear carrier, for reasons given previously. Therefore, the sear nose is not depressed and hence the safe position. Change over to the automatic position and pull the trigger; the head of the connector is raised and held in the vertical position, thus depressing the sear nose and holding it in this position, which obviously gives automatic fire as long as there are cartridges in the magazine. The tongue on the change lever tends to hold the connector vertically and the ramp on the sear carrier tends to cam the connector forward. The forces exerted by these two parts on the connector are opposed, hence the trigger mechanism is locked when the trigger has been pulled enough to release slide.

n. Set the change lever for semiautomatic fire. Pull the trigger slowly. At first the head of the connector rises and thereby depresses the sear nose, allowing the slide to go forward. If the press on the trigger is continued, the previously mentioned cam surface on the connector comes in contact with the cam surface of the sear carrier and the head of the connector is cammed forward against the pressure of the center leaf of the sear spring. The connector disengages the front arm of the sear and the two outside leaves of the sear spring depress it. The sear nose is thereby raised up in the path of the slide and engages the sear notch when the slide moves back, thus allowing only one shot to be fired. When the trigger is released, the center leaf of the sear spring presses the head of the connector downward and back under the forward end of the sear so that when the trigger is pulled again the action is repeated and a single shot is fired.

o. In the semiautomatic position the connector stop prevents the head of the connector being tipped so far forward that the sear spring cannot push it back in place when the trigger is released. The only function of the change lever in the semiautomatic position is the limiting of the upward travel of the trigger when its upper rear shoulders strike the top of the slot in the change lever, which in this position is turned down.

SECTION V

OPERATION

■ 25. OBJECT.—This section is designed to give the soldier instruction necessary for the operation of the rifle.

■ 26. WHEN TAKEN UP.—The operation of the rifle will be taken up at any convenient time after instruction in care and cleaning (sec. III) has been completed.

■ 27. USE OF DUMMY CARTRIDGES.—As prescribed in paragraph 18.

■ 28. TO LOAD MAGAZINE.—To load the magazine, place the wide end of the magazine filler over the top of the magazine so that the groove in the magazine filler fits over the catch rib of the magazine. Hold the magazine in the same relative position that it occupies in the rifle, that is, with the catch rib toward the operator. Then insert a clip of cartridges in the guides provided in the filler, and with the right thumb near the base, push the cartridges into the magazine. Each magazine will hold 20 rounds. (See fig. 30.)

■ 29. TO LOAD RIFLE.—Press the magazine release. Withdraw the empty magazine. Hold a loaded magazine with its base in the palm of the right hand, cartridges pointing to the front. Insert the magazine between the sides of the receiver

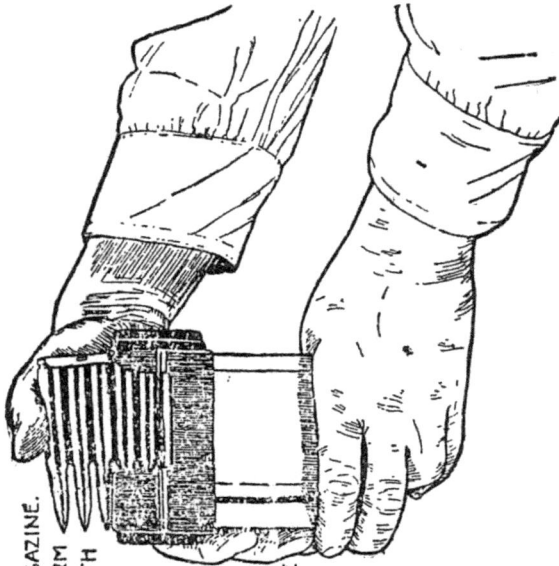

FIGURE 30.—To load magazine.

1. PLACE FILLER OVER MAGAZINE.

2. REST MAGAZINE ON FIRM BASE AND STEADY WITH LEFT HAND.

3. SEAT CLIP IN FILLER AND PUSH DOWN WITH THUMB AS IF LOADING RIFLE.

4. MAGAZINE SHOULD BE LOADED ONLY AS ABOVE OR WITH THE LOADING MACHINE.

in front of the trigger guard, and push it home smartly with the right hand.   The magazine can be inserted with the mechanism in either the cocked or forward position.   It is, however, ordinarily inserted after the rifle has been cocked.

■ 30. To UNLOAD RIFLE.—Press the magazine release and withdraw the magazine.   Let the bolt go forward by pulling the trigger.

■ 31. To FIRE RIFLE.—Press the trigger for each shot in semi-automatic fire.

■ 32. To Set Change Lever Control.—*a.* For semiautomatic fire, or single shot, push the change lever to the forward position, marked *F.*

b. For full automatic fire or continuous fire to the capacity of the magazine, set the change lever in the vertical position against the change lever stop, marked *A.*

c. To set the rifle at "safe," depress the change lever stop and pull the change lever rearward until it covers the change lever stop. This position is marked *S.* The change stop prevents the accidental setting of the change lever at safe, and at the same time allows a quick change from safe to either full automatic or semiautomatic fire.

■ 33. Gas Adjustment.—*a. General.*—(1) The rifle should normally be operated on the smallest port, and this setting will never be varied unless the rifle shows signs of insufficient gas. To aline the smallest port, screw in the gas cylinder with the combination tool until the shoulder of the gas cylinder is about one turn from the corresponding shoulder of the gas cylinder tube and the smallest circle on the cylinder head is toward the barrel. Lock the cylinder in position. If, upon firing, the rifle shows signs of insufficient gas, try setting the cylinder one complete turn on each side of the original setting. As soon as the proper setting has been obtained the rifleman will carefully note the position so that he can quickly assemble the cylinder to the proper point without trial.

(2) The larger ports are provided for use in case the action of the rifle has been made sluggish through the collection of dirt and grit, or the lack of oil under conditions which render prompt correction impracticable. For this reason the threads should be kept clean and oiled and the cylinder free to turn. The extractor, ejector, and the chamber of the barrel should be examined and cleaned and defects corrected when possible. Under adverse conditions, and when signs of insufficient gas become apparent, the cylinder should be unscrewed one-third of a turn, thus registering the medium circle and alining the medium port with the gas orifice. Repeat this operation in order to connect the largest port with the barrel.

(3) Excessive friction or dirt may sometimes prevent the complete forward movement of the bolt. This condition may also be caused by the recoil springs having become permanently set or short through continued use while excessively hot; in such cases, replace the recoil spring.

b. Results of insufficient gas.—(1) Failure to recoil (usually due to misalined or excessively clogged gas port, or extremely dirty mechanism).

(2) Failure to eject.

(3) Weak ejection.

(4) Uncontrolled automatic fire (exceptional).

c. Results of too much gas.—(1) Excessive speed, causing pounding.

(2) Excessive heat in gas operating mechanism.

■ 34. SAFETY PRECAUTIONS.—*a.* Automatic rifles will not be loaded except when on the firing line and with the muzzle pointed in the direction of the target.

b. Automatic rifles will not be carried loaded except in the presence of an enemy or a simulated enemy. Loaded rifles will be carried with the muzzle elevated or to the front.

c. Automatic rifles will be carried with the bolt forward at all times except in the presence of an actual enemy.

d. Automatic rifles will be assumed to be loaded whenever a magazine is in the receiver.

e. Never leave a patch, plug, or other obstruction in the muzzle or bore.

f. On the range, rifles are *cleared* before leaving the firing line. The automatic rifle is cleared by removing the magazine and releasing the bolt to its forward position.

SECTION VI

IMMEDIATE ACTION AND STOPPAGES

■ 35. OBJECT.—This section is designed to provide necessary instruction in the related subjects of immediate action and stoppages.

■ 36. WHEN TAKEN UP.—Instruction in immediate action and stoppages will be completed before any firing is done by the individual.

■ 37. IMMEDIATE ACTION.—*a. General.*—Immediate action is the unhesitating application of a probable remedy for a stoppage. Immediate action deals with the method of reducing stoppages and not the cause. It is taught as an unhesitating manual operation to be applied to reduce stoppages without detailed consideration of their causes.

b. Rifle fails to fire.—Pull the operating handle completely to the rear and then push it forward. Tap the magazine fully home. Aim and fire. If stoppage recurs, pull back the operating handle slowly to determine position of stoppages, remove the magazine, and apply proper remedy as explained in paragraph 38*d.*

■ 38. STOPPAGES.—*a. General.*—While immediate action and stoppages are closely related as to subject matter, the former is treated separately to emphasize its importance as an automatic and definite procedure to be applied to overcome stoppages. Proper care of the rifle before, during, and after firing will almost always eliminate stoppages. Stoppages which cannot be remedied by the application of immediate action can best be eliminated if the soldier has an understanding of the functioning of the weapon, and the causes of stoppages.

b. Types.—(1) Temporary stoppages have been divided into those found in four positions, dependent upon the position where the bolt stops. The position of the stoppage is determined by pulling the operating handle to the rear until it strikes the hammer pin.

(2) Boundaries of the positions are—

(*a*) *First position.*—Mechanism and operating handle fully closed.

(*b*) *Second position.*—Operating handle strikes hammer pin anywhere from fully closed to a point where operating handle plunger pin rides over raised shoulders on ribs of operating handle guideway.

(*c*) *Third position.*—Operating handle strikes hammer pin anywhere from second position to a point directly over *F* of change lever setting.

(*d*) *Fourth position.*—Operating handle strikes hammer pin anywhere between third position and rear.

59

c. Probable causes of stoppages.—(1) *Failure to extract.*
(*a*) Defective extractor.
(*b*) Dirt under extractor.
(*c*) Dirt in chamber.
(*d*) Pitted chamber.
(*e*) Weak extractor spring.
(*f*) Defective ammunition.
(2) *Failure to eject.*
(*a*) Insufficient gas.
(*b*) Defective extractor.
(*c*) Dirt under extractor.
(*d*) Ejector does not fit up close to bolt.
(*e*) Ejector binds on bolt.
(*f*) Ejector has too much backward play.
(*g*) Ejector bent backward or otherwise defective.
(*h*) Weak extractor spring.
(*i*) Defective ammunition.
(3) *Failure to breech.*
(*a*) Dirt between bolt and rear end of barrel.
(*b*) Primer on mechanism, generally in front of bolt.
(*c*) Defective bolt lock or pin.
(*d*) Defective magazine.
(*e*) Piston binding.
(*f*) Excessive friction.
(*g*) Recoil spring too short.
(4) *Insufficient gas.*
(*a*) Ports clogged.
(*b*) Poor fit between gas cylinder tube and bracket.
(*c*) Gas leakage around piston (worn cylinder).
(*d*) Piston binding, or cylinder dirty.
(*e*) Gas cylinder threaded in too far, or vice versa.
(5) *Ruptured cartridges.*
(*a*) Locking surface of bolt lock worn.
(*b*) Bearing between bolt and bolt lock worn.
(*c*) Face of bolt worn.
(*d*) Chamber of barrel worn or pitted.
(*e*) Locking shoulder of receiver worn.
(*f*) Bolt supports loose or worn.
(*g*) Defective ammunition.

d. Reduction of stoppages.—(1) *First-position stoppages.*—If the stoppage is in the first position, pull operating handle all the way back and watch ejection.

(*a*) If nothing is ejected, change magazine.

(*b*) If loaded round is ejected, change firing pin.

(*c*) If empty cartridge case is ejected, examine the correct adjustment of gas cylinder, turning to the next larger port if necessary.

(2) *Second-position stoppage.*—If the stoppage is in the second position, feel for obstruction or bur on face of bolt, in rear end of chamber, in bolt lock recess, or on bolt lock.

(3) *Third-position stoppages.*—If the stoppage is in the third position—

(*a*) If stoppage is due to a ruptured cartridge, use ruptured cartridge extractor to remove front part of ruptured case from the chamber. If no ruptured cartridge extractor is available, oil and sand nose of bullet, put it in chamber, and let bolt go forward. Pull back operating handle. This will usually extract the ruptured cartridge. Clean and oil chamber after doing this.

(*b*) If stoppage is not due to a ruptured cartridge, examine face of bolt for obstruction.

(4) *Fourth-position stoppages.*—If the stoppage is in the fourth position—

(*a*) If slide moved after trigger was pulled, use cleaning rod to push cartridge case from the chamber. If this stoppage recurs, clean ammunition and clean and lightly oil chamber with a rag.

(*b*) If trigger cannot be pulled or if the slide does not move when trigger is pulled—

    1. See if change lever is set on "Safe."

    2. Take out trigger guard and correct fault in the trigger mechanism.

e. The following table may be found of value. It includes stoppages outlined above and others that have not been covered but which may occur.

TABLE OF STOPPAGES

Position	Stoppage	Cause	Remedy in field
*First position*_____ (operating handle fully home).	1. Failure to feed_	1. Magazine troubles.	
		a. Magazine not fully home.	1. a. Push magazine home.
		b. Obstruction between lips of magazine a n d top cartridge.	b. Change magazine.
		c. Weak magazine spring.	c. Change magazine.
		d. Magazine dirty_	d. Change magazine; clean later.
		e. Magazine tube or lips dented or bent.	e. Change magazine.
		f. Magazine catch notch worn.	f. Change magazine.
	2. Failure to fire_	2. a. Broken or short firing pin.	2. a. Change firing pin.
		b. Weak recoil spring.	b. Change recoil spring.
		c. Excessive friction.	c. Clean and oil friction surfaces and chamber.
		d. Faulty ammunition — defective primers or charges.	d. Discard ammunition.
	3. Insufficient gas.	3. a. Gas cylinder not properly adjusted.	3. a. Correct adjustment of gas cylinder.
		b. G a s p o r t s clogged.	b. Turn cylinder to next larger port. Clean at first opportunity.
		c. Piston binding —dirty piston and cylinder.	c. Turn cylinder to next larger port. Clean and oil at first opportunity.
		d. Dirty chamber__	d. Turn cylinder to next larger port. Oil chamber. Clean and oil chamber at first opportunity.
		e. Lack of oil _____	e. Oil chamber and friction surfaces.

Table of Stoppages—Continued

Position	Stoppage	Cause	Remedy in field
*Second position*_____ (operating handle strikes hammer pin anywhere from fully closed back to top of raised shoulders on operating handle guideway).	1. Obstruction__	1. Extraneous matter or burs— *a*. On face of the bolt. *b*. In breech recess where bolt and receiver join. *c*. On bolt lock. *d*. In bolt lock recess.	1. Feel on face of bolt, in receiver and chamber, on bolt lock, and in bolt-lock recess for burs or extraneous matter. Remove extraneous matter or burs.
	2. Faulty ammunition.	2. Battered round.	
*Third position*_____ (operating handle strikes hammer pin between second position and point directly over *F* on receiver).	1. Ruptured cartridge.	1. Excessive head space.	1. Use ruptured cartridge extractor to remove ruptured cartridge case. If stoppage recurs, clean and oil chamber.
	2. Failure to feed completely.	2. *a*. Broken firing pin protruding from face of bolt. *b*. Other obstruction that prevents base of cartridge from sliding up across face of bolt.	2. *a*. Replace firing pin. *b*. Remove obstruction.
	3. Mechanism wedged tightly.	3. Obstruction— extraneous matter between bolt support and bolt lock.	3. Remove obstruction.
	4. Faulty ammunition.	4. Battered round.	
*Fourth position*_____ (operating handle strikes hammer pin betwee nthird position and all the way to the rear)	1. Failure to extract.	1. *a*. Dirty, rusted, or pitted chamber. *b*. Dirt under extractor.	1. Use cleaning rod to remove empty cartridge case and then— *a*. Clean and oil chamber. *b*. Clean face of bolt and extractor.

Position	Stoppage	Cause	Remedy in field
Fourth position— *Continued.*		c. Defective extractor.	c. Change extractor.
		d. Defective extractor spring.	d. Change extractor spring.
		e. Defective ammunition—soft rims on cartridges.	
	2. Trigger will n o t release slide, the piece being cocked.	2. Trouble in trigger mechanism—	
		a. Change lever set on S.	2. a. Set change lever on F or A.
		b. Improper assembly which results in failure of center prong of sear spring to push connector b a c k under front end of sear.	b. Take out trigger guard and examine. Replace necessary parts a n d assemble properly.
		c. Defective sear spring.	c. Same as b above.
		d. Defective or lost connector.	d. Same as b above.
	3. Obstruction.	3. Extraneous matter between ejector and bolt.	3. Remove obstruction.

f. Other stoppages.—In the event of stoppages that are not mentioned above and that cannot be reduced, the rifle should be turned in for examination and repair.

SECTION VII

SPARE PARTS AND ACCESSORIES

■ 39. SPARE PARTS.—*a.* The parts of any rifle will in time become unserviceable through breakage or wear resulting from continuous usage. For this reason spare parts are provided for replacement of the parts most likely to fail, for use in making minor repairs, and for general upkeep of the rifle.

Twenty-round magazines are also issued as spares, the quantity being based on the allowance of ammunition authorized. Sets of spare parts should be maintained as complete as possible at all times, and should be kept clean and lightly oiled to prevent rust. Whenever a spare part is used to replace a defective part in the rifle, the defective part should be repaired or a new one substituted in the spare-parts set. Parts that are carried complete should at all times be correctly assembled and ready for immediate insertion in the rifle. The allowances of spare parts and of 20-round magazines are prescribed in SNL A–4.

b. With the exception of replacements with the spare parts mentioned above, repairs or alterations by the using organizations are prohibited.

■ 40. ACCESSORIES.—*a. General.*—Accessories include the tools required for disassembling and assembling, and for the cleaning and preservation of the rifle. They must not be used for any purpose other than as prescribed. There are a number of accessories, the names or general characteristics of which indicate their uses or application. Therefore, detailed description or methods of use of such items are not outlined herein. However, accessories embodying special features or having special uses are described in the following subparagraphs.

b. Brush and thong, caliber .30, complete.—This consists of the brush, the tip, the weight, and the cord. The thong weight and tip are made of brass and are provided with holes in which the thong cord is knotted. The tip is provided with a cleaning patch slot and has threads provided on the end to receive the brush.

c. Brush, chamber-cleaning, M1.—The chamber-cleaning brush consists of a curved, flat, steel handle to which is hinged a chamber cleaning brush at one end and a small bristle dusting brush at the other end.

d. Brush, cleaning, caliber .30, M2.—The brush consists of a brass wire core with bristles and tip. The core is twisted in a spiral and holds the bronze bristles in place. The brass tip, which is threaded for attaching the brush to the cleaning rod, is soldered to the end of the core.

e. Case, accessory and spare parts, M1918.—This is a leather box-shaped case, approximately 2¼ inches wide, 3½ inches high, and 5½ inches long. It is used to carry the spare parts and a number of the smaller accessories.

f. Case, carrying, automatic rifle.—The carrying case for the automatic rifle is made of olive-drab cotton duck and has reinforced muzzle and breech ends. An olive-drab cotton webbing carrying strap is secured to the case by web chapes.

g. Case, cleaning rod, M1.—The case is a fabric container with five pockets, four of which hold the sections of the jointed cleaning rod, M1, while the fifth holds the cleaning brush, caliber .30, M2. The contents are secured in their pockets by a web billet and chape with buckle.

h. Extractor, ruptured cartridge, Mk. II.—The ruptured cartridge extractor has the general form of a caliber .30 cartridge. It consists of three parts—the spindle, the head, and the sleeve. To use the ruptured cartridge extractor, the live cartridges must be removed from the rifle. The ruptured cartridge extractor is then inserted through the ruptured opening of the case and pushed forward into the chamber. The bolt is let forward without excessive shock so that the extractor of the rifle engages the ruptured cartridge extractor. As the operating handle is drawn back the ruptured cartridge extractor holding the ruptured cartridge on its sleeve is extracted.

i. Filler, magazine.—The magazine filler is a pressed-steel adapter which may be fitted over the top of an empty magazine when loading. Its method of use is shown in figure 30.

j. Rod, cleaning, M2.—This is a straight rod consisting of two sections permanently fastened together with a swivel joint. The front end has a threaded hole for attaching the cleaning brush and a slot for holding a cleaning patch. The rear end is provided with a tubular steel handle which swivels on the rod.

k. Rod, cleaning, jointed, M1.—This steel rod consists of five sections, the first two of which are permanently fastened together with a swivel joint. The first section has a slot formed for holding a cleaning patch and a threaded hole for attaching the cleaning brush. The rear section is provided with a tubular steel handle which swivels on the rod.

l. Sling, gun, M1907.—The gun sling is fastened to the swivels provided on the rifle. It consists of a long and short strap, either of which may be lengthened or shortened to suit the particular soldier using it.

m. Tool, cleaning, gas cylinder.—This is a special tool for cleaning the gas-operating mechanism. The ends of the tool body may be used to scrape carbon from the interior of the gas cylinder and from the face of the gas piston. The drift which is attached to the body may be used to remove carbon deposits from the gas ports and from the grooves of the gas piston. The carbon must be completely removed, but care also must be exercised to avoid scoring or damaging the gas cylinder walls or the grooves of the gas piston.

n. Tool, combination.—This tool consists of a steel body having two spanner wrenches and two screw-driver ends. The small spanner is used to turn the gas cylinder and flash hider, the large spanner the rifle barrel. The small screw driver at the end of the large spanner may be used for the removal of small screws, the larger screw driver for the removal of the buttstock bolt and the forearm screws.

o. Accessories for pack transport.—Additional accessories are provided for pack transport of the automatic rifle. They include the rifle hangar, the carrying case, ammunition chests, and several straps used for securing the matériel transported.

SECTION VIII

AMMUNITION

■ 41. GENERAL.—The information in this section pertaining to the several types of cartridges authorized for use in the Browning automatic rifle, caliber .30, M1918 and M1918A1, includes a description of the cartridges, means of identification, care, use, and ballistic data.

■ 42. CLASSIFICATION.—Based upon use, the principal classifications of ammunition for this rifle are—

a. Ball.—For use against personnel and light matériel targets.

b. Tracer.—For observation of fire and incendiary purposes.

c. Armor-piercing.—For use against armored vehicles, concrete shelters, and similar targets.

d. Dummy.—For training.   (Cartridges are inert.)

■ 43. AMMUNITION LOT NUMBER.—When ammunition is manufactured, an ammunition lot number, which becomes an essential part of the marking, is assigned in accordance with specifications.   This lot number is marked on all packing containers and the identification card inclosed in each packing box.   It is required for all purposes of record, including grading and use, reports on condition, functioning, and accidents, in which the ammunition might be involved.   Since it is impracticable to mark the ammunition lot number on each individual cartridge, every effort should be made to maintain the ammunition lot number with the cartridges once they are removed from their original packing.   Cartridges which have been removed from the original packing and for which the ammunition lot number has been lost are placed in grade 3.   It is therefore necessary, when cartridges are removed from original packings, that they be so marked that the ammunition lot number be preserved.

■ 44. GRADE.—Current grades of existing lots of small arms ammunition are established by the Chief of Ordnance, and are published in Ordnance Field Service Bulletin No. 3–5. No lot other than that of current grade appropriate for the weapon will be fired.   *Grade 3 ammunition is unserviceable and will not be fired.*

■ 45. IDENTIFICATION.—*a. Markings.*—The contents of original boxes are readily identified by the markings on the box.   Similar markings on the carton label identify the contents of each carton.

b. Color bands.—Color bands, painted on the sides and ends of the packing boxes, further identify the various types of ammunition.   The following color bands are used:

Cartridge, armor-piercing	Blue on yellow.
Cartridge, ball	Red.
Cartridge, tracer	Green on yellow.
Cartridge, dummy	Green.

c. Types and models.—(1) The following types and models of caliber .30 ammunition are authorized for use in the Browning automatic rifle, caliber .30, M1918 and M1918A1:

(*a*) Cartridge, ball, caliber .30, M2.

(*b*) Cartridge, ball, caliber .30, M1.

(*c*) Cartridge, armor-piercing, caliber .30, M1.

(*d*) Cartridge, armor-piercing, caliber .30, M2.

(*e*) Cartridge, armor-piercing, caliber .30, M1922.

(*f*) Cartridge, dummy, range, caliber .30, M1 or M1921.

(*g*) Cartridge, tracer, caliber .30, M1.

When removed from their original packing containers, the cartridges may be identified except as to ammunition lot number and grade by physical characteristics described below.

(2) *Armor-piercing.*—All models of caliber .30 armor-piercing ammunition are distinguished by the nose of the bullet, which is painted black for a distance of approximately one-fourth of an inch from the tip. The bullets have gilding-metal jackets.

(3) *Ball.*—All models of caliber .30 ball ammunition, except the M1906, have bullets with gilding-metal jackets. The jacket of the M1906 bullet is cupronickel, which has a silvery appearance. The gilding-metal jacket of the M2 bullet is tin-coated and hence resembles the M1906 bullet in appearance. The gilding-metal jacket of the M1 bullet is copper-colored.

(4) *Tracer.*—Caliber .30 tracer ammunition may be identified by the nose of the bullet, which is painted red for a distance of approximately one-fourth of an inch from the tip.

(5) *Dummy.*—The caliber .30 corrugated dummy cartridge may be identified by the corrugations formed in the cartridge case.

■ 46. CARE, HANDLING, AND PRESERVATION.—*a.* Small-arms ammunition is not dangerous to handle. Care, however, must be exercised to keep the boxes from becoming broken or damaged. All broken boxes must be immediately repaired and all original markings transferred to the new parts of the box. The metal liner should be air-tested and sealed if equipment for this work is available.

b. Ammunition boxes should not be opened until the ammunition is required for use. Ammunition removed from

69

the airtight container, particularly in damp climates, is apt
to corrode, thereby causing the ammunition to become unserviceable.

c. The ammunition should be protected from mud, sand,
dirt, and water.   If it gets wet or dirty, wipe it off at once.
Light corrosion, if it forms on cartridges, should be wiped
off.   However, cartridges should not be polished to make
them look better or brighter.

d. No caliber .30 ammunition will be fired until it has been
identified by ammunition lot number and grade.

e. Do not allow the ammunition to be exposed to the direct rays of the sun for any length of time.   This is likely
to seriously affect its firing qualities.

■ 47. STORAGE.—Whenever practicable, small-arms ammunition should be stored under cover.   Should it be necessary to
leave small-arms ammunition in the open, it should be raised
on dunnage at least 6 inches from the ground and the pile
covered with a double thickness of paulin.   Suitable trenches
should be dug to prevent water from flowing under the pile.

■ 48. BALLISTIC DATA.—Approximate maximum ranges are
as follows:

	Yards
Cartridge, ball, caliber .30, M2	3,450
Cartridge, ball, caliber .30, M1	5,500
Cartridge, armor-piercing, caliber .30, M1	4,000
Cartridge, armor-piercing, caliber .30, M2	4,500
Cartridge, armor-piercing, caliber .30, M1922	4,400
Cartridge, tracer, caliber .30, M1	3,450

CHAPTER 2

MARKSMANSHIP—KNOWN-DISTANCE TARGETS

		Paragraphs
Section I.	General	49–51
II.	Preparatory markmanship training	52–75
III.	Courses to be fired	76–78
IV.	Range practice	79–88
V.	Regulations governing record practice	89–119
VI.	Targets and ranges	120–121

SECTION I

GENERAL

■ 49. OBJECT.—The object of this chapter is to provide a thorough and uniform method of training individuals to be good shots and of testing their proficiency in firing at known-distance targets with the Browning automatic rifle, caliber .30, M1918, without bipod.

■ 50. FUNDAMENTALS.—To become a good automatic rifle shot the soldier must be trained in the following essentials of good shooting:

a. Correct sighting and aiming.

b. Correct positions.

c. Correct trigger squeeze.

d. Correct application of rapid fire principles.

e. Knowledge of proper sight adjustments.

■ 51. PHASES OF TRAINING.—*a.* Marksmanship training is divided into the following phases:

(1) Preparatory marksmanship training.

(2) Range practice.

b. No individual will be given range practice until he has had a thorough course in preparatory training.

c. The soldier will be proficient in mechanical training before he receives instruction in marksmanship training.

d. Every man who is to fire on the range will be put through the preparatory course regardless of previous qualifications.

SECTION II

PREPARATORY MARKSMANSHIP TRAINING

■ 52. GENERAL.—*a*. The purpose of preparatory marksmanship training is to teach the soldier the essentials of good shooting and to develop fixed and correct shooting habits before he undertakes range practice.

b. Preparatory marksmanship training is divided into six steps, as follows:

(1) Sighting and aiming exercises.

(2) Position exercises.

(3) Trigger-squeeze exercises.

(4) Rapid-fire exercises.

(5) Instruction in the effect of wind, sight changes, and use of the score book.

(6) Examination of men before starting range practice.

■ 53. WHEN TAKEN UP.—Preparatory marksmanship will be taken up in the period stated in training programs and will precede range practice. Preparatory marksmanship training can be covered in 2 training days.

■ 54. EQUIPMENT.—*a. List of equipment for each eight-man group.*

(1) Two sighting bars.

(2) Two automatic rifles and rests.

(3) Two 3-inch sighting disks. (See fig. 31.)

(4) Four small aiming targets.

(5) One long-range sighting disk.

(6) Two small boxes.

(7) One target frame covered with blank paper for long-range triangles.

(8) One score book for each man.

(9) Two blank examination forms as shown in paragraph 75.

(10) One rapid-fire target with curtain for each three groups.

b. Preparation of equipment.—(1) *Sighting bar.*—Construct the sighting bars from trim lumber and tin strip to the dimensions and design shown in figure 31. The sighting bars, as all other equipment, should be constructed so as to

72

present a neat appearance. The tops of the sighting bars, their front and rear sights, and their eyepieces are painted black.

(2) *Automatic rifle and rest.*—An empty ammunition box or any similar box with notches cut in the ends to fit the automatic rifle closely makes a good automatic rifle rest. The automatic rifle is placed in those notches with the trigger guard just outside one end. The sling is loosened and pulled to one side. The box is half-filled with earth or sand to give it stability.

(3) *Sighting disks.*—Sighting disks are of two sizes. The disk to be used at a distance of 50 feet is shown in figure 31.

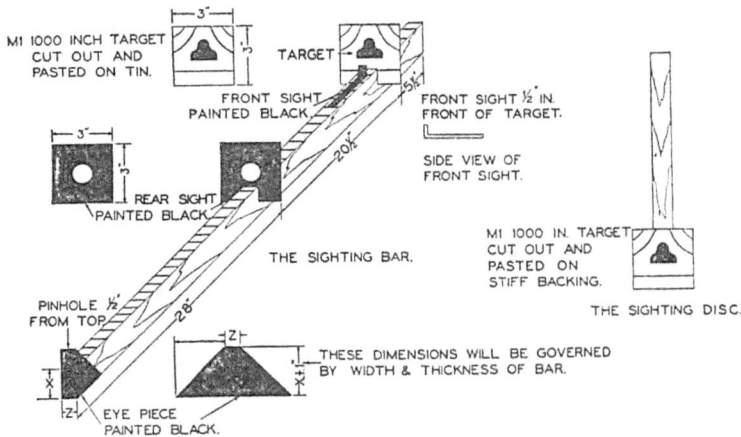

FIGURE 31.—Construction of sighting bar and sighting disk.

The disk to be used at 200 yards is constructed by pasting the black silhouette of a standard D target on some stiff backing, and attaching a 4-foot handle. The sighting disks have holes in their centers of a size sufficient to admit the point of a pencil.

(4) *Blackening sights.*—In all preparatory exercises involving aiming and in all range firing both sights of the rifle are blackened. Before blackening, the sights are cleaned and all traces of oil removed. The blackening is done by holding each sight for a few seconds in the point of a small flame which is of such a nature that a uniform coating of

lampblack will be deposited on the metal. Materials commonly used for this purpose include carbide or kerosene lamps, candles, small pine sticks, and shoe paste.

■ 55. DUTIES OF LEADERS.—*a. Battalion commander.*—He requires the officers and noncommissioned officers to be familiar with the prescribed methods of instruction and coaching, supervises the instruction within his battalion, and requires the companies to follow the preparatory exercises and methods of coaching carefully and in detail.

b. Company commander.—He requires the prescribed methods of instruction and coaching to be carried out carefully and in detail within his company and supervises and directs the platoon leaders.

c. Platoon leader.—He supervises and directs the squad leaders in training their squads and examines each man in his platoon as required in paragraph 75. He keeps up a copy of the form shown in paragraph 75 for each man.

d. Sergeants.—They assist in the instruction and perform any other duties as directed by the company and platoon commanders. They may assist the platoon leaders in keeping the forms referred to in *c* above.

e. Squad leaders.—(1) He sees that each man in his squad is occupied in the designated preparatory training.

(2) He keeps up a separate copy of the form shown in paragraph 75, and promptly enters the grades made by his men as the work progresses. He has this form ready for the platoon leader's inspection at any time.

(3) He requires the coaches to correct errors.

NOTE.—The duties of company commanders and squad leaders in this paragraph apply also to battery and troop commanders and section leaders of units other than Infantry.

■ 56. METHOD OF INSTRUCTION.—*a.* Men are grouped in pairs, as coach and pupil, and alternate in assisting and coaching each other.

b. Correct shooting habits are developed during the preparatory exercises, and to this end the careful execution of details is required. Training proceeds expeditiously to maintain interest. Care will be taken to avoid holding the men in position until they become uncomfortable. Frequent short rests will be given.

c. Arrangements are made to enable officers and noncommissioned officer instructors to complete their own preparatory marksmanship training prior to that of their men in order that they may give their entire attention to the men whom they are to direct and instruct.

■ 57. DUTIES OF COACHES.—The successful conduct of the preparatory exercises largely depends upon the attention of the coaches to their duties. Officers and noncommissioned officers are specifically charged with the supervision of coaches as well as of pupils. They will require the coaches to have their pupils execute all steps of the preparatory exercises correctly. The duties of a coach are specific and during the progress of the preparatory exercises include necessary correction of the pupil to see that—

a. Sights are blackened.

b. Gun sling is properly adjusted.

c. Position is taken correctly.

d. Slack is taken up promptly.

e. Aim is carefully taken.

f. Breath is held during aiming (by watching pupil's back).

g. Trigger is pressed properly.

h. Pupil calls the shot.

■ 58. FIRST SIGHTING AND AIMING EXERCISE—SIGHTING BAR.—
The instructor or squad leader shows a sighting bar to his group and explains its use as follows, being careful to point out the various parts of the bar as he refers to them:

a. The front and rear sights on the sighting bar represent enlarged rifle sights.

b. The eyepiece on the sighting bar has no counterpart on the rifle. The eyepiece on the sighting bar is used as an aid to instruction because it enables the alinement of the sights to be demonstrated easily. The movable target on the sighting bar enables any alinement of the sight upon the silhouette to be shown.

c. He next explains the peep sight to his group and shows each man a correct sight alinement with the targets removed. (See fig. 32.)

d. He next describes the correct aim. He explains that the top of the front sight is centered on the rear sight so as just to touch the bottom of the silhouette.

e. He explains that the eye should be focused on the sight picture, being sure that his front sight is distinct against the target, and he assures himself by questioning the pupils that each man understands what this means.

f. He adjusts the rear sight of the sighting bar and the movable target so as to illustrate the correct aim and has each man observe it by looking through the eyepiece.

g. He adjusts the rear sight and the movable target of the sighting bar so as to illustrate various small errors and has each man of the group detect and describe them.

FIGURE 32.—Correct sight with sighting bar.

h. Each man will then again be shown the bar with the correct aim illustrated.

i. Each man will then be required to adjust the sighting bar with the correct aim until he is proficient, the coach and pupil method being used.

■ 59. SECOND SIGHTING AND AIMING EXERCISE—ALINING SIL-HOUETTE AND SIGHTS.—A rifle for each subgroup is placed in a rifle rest and pointed at a blank sheet of paper mounted on a box at which a soldier with the small disk is stationed as marker. The coach or an instructor takes the prone position and without touching the rifle looks through the sights. He directs the marker, by voice or signal, to move the small disk

FIGURE 33.—Position for second sighting and aiming exercise. ①

②

FIGURE 33.—Position for second sighting and aiming exercise—Continued.

until the bottom of the silhouette is in correct alinement with the sights. He then calls, "Hold," at which signal the marker will hold the small disk in position. The coach or instructor moves away from the rifle and directs the pupil to look through the sights in order to observe the correct aim. He then requires the pupil to execute the exercise for himself, being careful to check the alinement which the pupil obtains.

■ 60. THIRD SIGHTING AND AIMING EXERCISE—MAKING SHOT GROUPS.—*a.* The object of this exercise is to teach uniform and correct aiming.

b. The exercise is conducted as follows: The rifle with blackened sights is placed in a rifle rest and pointed at a blank sheet of paper mounted on a box 50 feet away. The pupil takes the prone position without touching the rifle or rests and looks through the sights. The pupil or coach directs the marker to move the small disk until the bottom of the silhouette is in correct alinement with the sights and then calls, "Hold." The instructor checks the alinement and then calls, "Mark." The marker immediately marks a dot on the paper with a sharp-pointed pencil inserted through the hole in the silhouette. The small disk is removed and the dot numbered. The pupil repeats this operation until three dots, numbered 1, 2, and 3, have been made. These dots outline the aiming group and the pupil's name is written under it. The size and shape of the aiming group will be discussed with the coach or instructor and the cause of error pointed out and corrected. This exercise is repeated until proficiency is attained. A good group of three marks can be covered by the eraser of an ordinary pencil (or a circle 0.2 inch in diameter).

c. A similar exercise will also be held during the period of preparatory marksmanship training at 200 yards with the movable silhouette. If the exercise is properly handled, it helps greatly to sustain interest in the work and to teach correct aiming. At 200 yards a man should be able to make an aiming group that can be covered with the small (3-inch) sighting disk.

d. Tracings are made of each man's 200-yard aiming group. These tracings are marked with the men's names, turned over to the platoon leader for his information, and shown to the men with appropriate mention of errors to be corrected.

e. The triangle exercise may be continued during the remaining periods of preparatory marksmanship training to maintain interest and to secure the proficiency of men who require special instruction.

■ 61. POSITION EXERCISES.—*a. General.*—Instruction in position will include the use of the gun sling, holding the breath, and aiming in each position. Small targets should be set up for each position to assist the aim.

b. Scope of instruction.—Detailed instruction will be given in each of the positions described in paragraphs 62 to 66, inclusive.

c. General rules governing all positions.—(1) To assume any position except the prone position, first half-face to the right and then assume the position.

(2) Upon assuming any position there is a point at which the rifle points naturally and without effort. If this point is not the center of the target, the whole body must be shifted so as to bring the target into proper alinement.

(3) The right hand grasps the small of the stock. The thumb may be around the small of the stock or on top of the stock.

(4) The left hand is not forced forward farther than is comfortable and convenient in the prone, kneeling, or sitting position. (In the prone position with muzzle rest, the left hand grasps the sling.) The left wrist is straight and the rifle is placed in the crotch formed by the thumb and index finger and resting on the heel of the hand.

(5) The left elbow is as nearly under the rifle as it can be placed without strain.

(6) The trigger is pressed with the second joint of the index finger; the first joint may be used if necessary.

(7) The cheek firmly rests against the stock and is placed as far forward as possible without strain to bring the eye near the rear sight.

(8) Men will not be permitted to shoot in the left-handed position.

d. Gun sling.—(1) The gun sling, properly adjusted, is of great assistance in shooting in that it helps to steady the rifle. When used, each man will be assisted by the instructor in securing the correct adjustment for his sling. In a firing position the sling is adjusted to give firm support without discomfort to the soldier. The gun sling is readjusted for drill purposes by means of the lower loop without changing the adjustment of the upper loop.

(2) There are two authorized adjustments—the *loop* sling and the *hasty* sling. The hasty sling is more rapidly adjusted than the loop sling but it gives less support.

(a) *Loop adjustment.*

    1. Loosen the lower loop.

    2. Insert the left arm through upper loop from right to left, so that the upper loop is near the shoulder and well above the biceps muscle.

    3. Pull the keepers and hook close against the arm to keep the upper loop in place.

    4. Move the left hand over the top of the sling and grasp the forearm of the rifle near the center so as to cause the sling to lie smoothly along the hand and wrist. The lower loop, not used in this adjustment, should be so loose as to prevent any pull upon it. Neither end will be removed from either swivel.

(b) *Hasty sling adjustment.*

    1. Loosen the lower loop.

    2. Grasp the forearm of the rifle near the center with the left hand and grasp the small of the stock with the right hand.

    3. Throw the sling to the left and catch it above the elbow and high on the arm.

    4. Remove the left hand from the rifle, pass the left hand under the sling, then over the sling, and regrasp the rifle with the left hand so as to cause the sling to lie along the hand and wrist. The sling may be given one-half turn to the left and then adjusted. This twisting causes the sling to lie smoothly along the hand and wrist.

e. Holding breath.—The breath is held during aiming. To accomplish this, draw a little more air into the lungs than is used in an ordinary breath. Let out a little of this air and hold the rest naturally and without constraint.

f. Taking up trigger slack.—The first movement of the trigger which takes place when light pressure is applied is called *taking up the slack*. It is part of the position exercise because this play must be taken up by the finger as soon as the correct position is assumed and before careful aiming is begun. The entire amount of slack is taken up by one positive movement of the finger.

g. Canting rifle.—In all positions the rifle is squarely held; that is, not tipped or canted from a vertical plane passing through its long axis. It should be understood, however, that unless it is pronounced this error in position will not materially affect the aim or the shot.

■ 62. PRONE POSITION WITH HASTY OR LOOP SLING.—In assuming the prone position the body should be as straight behind the piece as the conformation of the firer will permit. The angle of the body to the line of aim should not exceed 15°. The legs should be well apart, the inside of the feet flat on the ground, or as nearly so as can be attained without strain. The left elbow should be approximately under the piece and the right elbow drawn in fairly close to the body. The position of the elbow should distribute the weight evenly and raise the chest slightly off the ground. The right hand grasps the small of the stock. The left hand should not be forced forward to the sling swivel, but should be as far forward as is comfortable and convenient for the individual firer. The rifle is placed in the crotch formed by the thumb and index finger and rests on the heel of the hand. The sling should be just sufficiently tight to offer support, but not so tight as to have a tendency to pull the left elbow to the left. The cheek should be resting firmly against the stock with the eye looking through the rear sight, without straining the neck muscles. The right thumb may be over the small of the stock, or on top of the stock; it should not be placed alongside the stock. The exact details of the position will vary, depending upon the conformation of the individual firer. However, the firer must secure a position that will not be changed by the recoil of the weapon. If the left elbow or the right elbow is moved by the recoil of the piece, the firer will find it necessary to "fight" his way back into position after each shot and valuable time will be lost. When the correct position has been attained it will be found that upon discharge the muzzle will move slightly up and very slightly to the right, and that it will then settle back close to the original aiming point.

■ 63. PRONE POSITION WITH SANDBAG REST.—The prone position as described in paragraph 62 is taken with the following exceptions:

a. The hasty sling is adjusted on the rifle but not used on the arm by the firer.

b. The left hand of the firer grasps the sling under the forearm of the rifle, back of the hand up, and is pulled down and to the rear.

c. The gas cylinder tube is supported by a sandbag. When first adjusted to the firer, the sandbag should be too high; that is, high enough so that the aim of the firer will not be lower than the top of the bull's-eye on the target. Further adjustment is made by pushing down on the barrel until the sights are alined on the proper place on the target. Any tendency to have the sandbag too low must be promptly corrected by the coach.

■ 64. SITTING POSITION.—The firer sits half faced to the right, feet well apart and well braced on the heels which are slightly dug into the ground, ankles relaxed, body leaning well forward from the hips with back straight, both arms resting inside the legs and well supported, cheek resting firmly against the stock and placed as far forward as possible without straining, left hand as far forward as convenient and comfortable, wrist straight, rifle placed in the crotch formed by the thumb and index finger and resting on the heel of the hand. In this position the feet may be slightly lower than the ground on which the firer sits. Sitting on a low sandbag is authorized. Necessary changes to adapt the position to the conformation of the man are authorized. Instruction in the sitting position is limited to that sufficient to acquaint the men with it, as the use of this position is regarded as exceptional.

■ 65. KNEELING POSITION.—The firer kneels half faced to the right on the right knee, sitting on the right heel, the left knee bent so that the lower left leg is vertical as viewed from the front, left arm well under the rifle and resting on the left

① With sandbag—side view.

FIGURE 34.—Prone position.

② With sandbag—front view.

FIGURE 34.—Prone position—Continued.

③ With loop sling.

FIGURE 34.—Prone position—Continued.

④ With hasty sling.

Figure 34.—Prone position—Continued.

87

FIGURE 35.—Sitting position.

FIGURE 36.—Kneeling position.

knee with the point of the elbow beyond the kneecap, right elbow approximately at the height of the shoulder, cheek resting firmly against the stock and placed as far forward as possible without strain. Sitting on the side of the foot instead of the heel is authorized. The center of balance of the firer should be low and forward.

■ 66. ASSAULT FIRE POSITION.—In this position, the automatic rifle is held with the butt under the right armpit; clasped

FIGURE 37.—Assault fire position.

firmly between the body and the upper portion of the arm, the sling over the left shoulder.

■ 67. PROCEDURE IN CONDUCTING POSITION EXERCISES.—*a*. Small bull's-eyes are used as aiming points. These bull's-eyes should be placed at a range of 1,000 inches and at different heights so that in aiming from various positions the automatic rifle will be nearly horizontal, or standard known-distance targets may be installed at distances used on the known-distance range.

b. Before taking up each phase of the position exercise, the instructor assembles his squad or group and—

(1) Shows the proper method of blackening the front and rear sights of the automatic rifle, and has each pupil blacken his sights.

(2) Explains and demonstrates the hasty sling adjustment and assists each pupil to adjust his sling. He explains the loop-sling adjustment and assists each pupil to adjust his sling.

(3) Explains and demonstrates the proper manner of taking up the trigger slack and has each pupil practice it.

(4) Explains and demonstrates the proper manner of holding the breath and has each pupil practice it.

(5) Explains the general rules which apply to all positions.

(6) Explains and demonstrates the different positions.

c. Following explanations and demonstrations the instruction becomes individual by the coach-and-pupil method. Each pupil, after seeing that his sights are blackened, adjusts his sling, takes position, takes up the slack, aims carefully, and holds his breath while aiming. As soon as his aim becomes unsteady, the exercise ceases. After a short rest the pupil repeats the exercise without further command. The trigger is not squeezed in the position exercises. Exercises are conducted in all positions.

d. Duties of coach.—In the position exercises the coach sees that—

(1) Sights are blackened.

(2) Gun sling is properly adjusted, is tight enough to give support, and is high up on the arm.

(3) Proper position is taken.

(4) Slack is taken up promptly.

(5) Pupil aims.

(6) Breath is held while aiming.

The coach checks the pupil's manner of holding his breath by watching his back.

■ 68. TRIGGER SQUEEZE EXERCISE.—*a. General.*—In both slow and time fire it is important to squeeze the trigger in such a way as to fire the rifle without affecting the aim. With the automatic rifle it is necessary to hold the trigger to the rear a brief interval after the shot is fired in order that two shots will not be fired automatically. A good shot knows that he cannot get results by jerking the trigger at the exact moment at which the sights are alined on the mark. He therefore holds his aim as steadily as possible and squeezes the trigger by a constantly increasing pressure applied directly to the rear. This method of squeezing the trigger must be carried out in all preparatory exercises or the value of the practice is lost.

b. Trigger squeeze exercises will be carried out in the preparatory exercises regardless of the fact that the men undergoing instruction may have just completed firing with the U. S. rifle, caliber .30, M1903, or the U. S. rifle, caliber .30, M1. The forward movement of the bolt when the trigger is squeezed is confusing to many men and causes them to allow the alinement of the sights to become incorrect. The piece must be held steady and in perfect alinement during this forward movement.

c. Calling shot.—The pupil must always notice where the sights are pointed at the instant the rifle is fired, or when the bolt reaches its forward position in simulated fire, and call out at once where he thinks the bullet will hit. In rapid fire he should call the last shot.

d. Exercises.—(1) The coach-and-pupil method being used, the pupil is first taught trigger squeeze in the prone position with sandbag. In this position he can hold steadily while he squeezes the trigger. After proficiency is obtained in the prone position with the sandbag, trigger squeeze will be practiced in the sitting and kneeling positions.

(2) In all exercises where fire is simulated, men will carry out the correct principles of aiming, squeezing the trigger, and calling the shot.

■ 69. Rapid-Fire Exercise.—*a.* Rapid-fire exercises enable the soldier to gain dexterity in the manipulation of the automatic rifle. Efficient manipulation is an important factor in automatic rifle firing.

b. Rapid-fire exercises are held at short ranges with D targets in frames as aiming points. The exercises include observance of the principles of sighting, positions, and trigger squeeze as taught in the preceding exercises.

c. During semiautomatic firing it is essential that the firer count his shots so that he will know when the last one has been fired and the magazine is empty. This prevents sending the bolt home on an empty chamber, which would necessitate pulling back the operating handle and so waste time.

d. When the last shot in the magazine is fired, the bolt being in the rearward position and the rifle held horizontally, the automatic riflemen pushes the magazine release with his right thumb and the magazine falls out of its own weight.

e. Magazines are so placed in the belt that when grasped and carried forward by the right hand the long portion will be to the rear. Thus they may be readily inserted in the magazine opening in the receiver.

f. Each soldier changes his own magazines. He must be able to do it in 2 to 4 seconds while in any position. He tests all magazines before firing to see that they will fall out of their own weight when empty.

g. During rapid fire preparatory exercises, the bolt being left in the forward position, and the change lever being set at *F,* firing is simulated by squeezing the trigger until it clicks for each shot.

h. Duties of coach.—In a rapid-fire exercise the coach insures that—

(1) Sights are set for the ranges designated and are blackened.

(2) Gun sling is properly adjusted.

(3) Correct position is taken.

(4) Slack is taken up promptly.

(5) Breath is held while aiming.

(6) Trigger is squeezed properly.

(7) Each man counts his shots.

(8) Magazines are dropped out after the last shot.

(9) New magazine is drawn from the belt and placed in the receiver with one rapid, smooth movement.

i. Empty magazines are picked up and placed in the belt. In known-distance firing they will be placed in the belt at the conclusion of the firing of the exercise.

Firing.

Pressing in Magazine Release with Right Thumb

Magazines Should be Sufficiently Smooth to Drop Out

New Magazine

Reaching for Old Magazine

Replacing Old Magazine in Belt Before Refiring

FIGURE 38.—Changing magazine drill.

■ 70. Sɪɢʜᴛ Sᴇᴛᴛɪɴɢ.—*a*. Instruction or a review of sight setting as set forth in the following paragraphs of this section will be completed during the period devoted to preparatory marksmanship training. Instruction or a review of the use of the scorebook in accordance with the instructions contained therein will be similarly completed. In connection with this subject those under instruction will be informed that care exercised in the required use of the scorebook during range practice will materially assist them in becoming qualified shots.

b. Practical instruction in sight setting and in keeping the scorebooks may be given indoors or under shelter during inclement weather with such simple equipment as blackboards, targets, the automatic rifle, scorebooks, and pencils.

■ 71. Eʟᴇᴠᴀᴛɪᴏɴ Rᴜʟᴇ.—*a*. Changing the elevation 100 yards at any range will give a change on the target, in inches, equal to the square of the range (expressed in hundreds of yards).

Example: At 200 yards, changing the elevation 100 yards makes 4 inches' change on the target; at 300 yards, 9 inches'; at 500 yards, 25 inches'; at 600 yards, 36 inches'. This rule is not exact, but near enough for all practical purposes.

b. The horizontal lines in the model targets in the scorebook also show how much change to make in the elevation at each range. When a change in elevation is necessary, it is best to consult the model target in the scorebook before deciding how much of a change to make.

■ 72. Wɪɴᴅᴀɢᴇ.—The horizontal clock system is used to describe the direction of the wind. In this system the firer is assumed to be at the center of a clock and the target at 12 o'clock. A 3-o'clock wind then blows directly from the right, a 9-o'clock wind directly from the left, and other winds from their corresponding directions on the clock. An aiming point is taken to the right to counteract the effects of winds coming from the right of the clock and to the left to counteract those coming from the left. The table of wind allowances shown in paragraph 8*c* on page 9 of the Individual Score Book for the Rifle (W. D., A. G. O. Form No. 82) is utilized to obtain the distance to aim off the silhouette for the first shot.

■ 73. EXPLANATION OF ZERO.—a. An explanation of the zero of the automatic rifle should be included in the instruction in sight setting. The zero of the automatic rifle for any range is that sight setting in elevation, and the aiming point, which will center the shot group on the target on a day when there is good light and no wind. It may vary for the same rifle with different men on account of the differences in eyesight. Each man should understand this explanation of the zero of a rifle, and that he will be required to keep a record of the zero elevations and aiming points for his own rifle for the various ranges in his score book.

b. Instructions for zeroing the rifle on the 1,000-inch range and on the known-distance range are given in section IV.

■ 74. SCORE BOOK.—a. The use by the soldier of a score book to keep a personal record of the results and conditions of his firing throughout the period of range practice is as necessary with the automatic rifle as with the service rifle. The Individual Score Book for the Rifle is used for the automatic rifle.

b. The use of the score book for the automatic rifle is the same as for the service rifle. Certain variations are noted as follows:

(1) The aiming point used for each shot is plotted in the column headed "Call."

(2) Upon completion of the score, no entry is made after the words "zero windage." In place of zero windage the aiming point for the rifle at that range is determined, and its location indicated by an *X* on the recording target.

■ 75. EXAMINATION.—a. Men will be examined prior to proceeding to range practice to determine their proficiency in the subjects covered in chapter 1 and in section II of chapter 2.

b. *Type examination on mechanical training.*—The following questions and answers may constitute the examination on mechanical training:

Q. What is the name of the rifle? A. The Browning automatic rifle, caliber .30, M1918.

Q. What type weapon is it? A. Shoulder weapon, gas-operated, air-cooled, and magazine-fed.

Q. How is the barrel cooled? A. There is no special cooling device. The barrel is merely exposed to the air.

Q. What is meant by gas operated? A. All of the functions of the automatic rifle, such as extraction and feeding, are accomplished by a small portion of gas escaping through a port in the barrel and impinging on a piston.

Q. How many rounds are carried in the magazine? A. Twenty.

Q. How much does the automatic rifle weigh with sling and without magazine? A. Fifteen pounds fourteen ounces.

Q. From what positions may the automatic rifle be fired? A. From any of the positions used with the service rifle.

Q. What is the best rate of fire? A. Semiautomatic fire at the rate of 40 to 60 shots per minute.

Q. What is meant by semiautomatic fire? A. Squeezing the trigger for each shot that is fired.

Q. How is the examination on disassembling and assembling of the automatic rifle conducted? A. (Man being examined disassembles and assembles the rifle, naming parts.)

Q. How is the examination on disassembling and assembling of the trigger mechanism conducted? A. (Man being examined disassembles and assembles trigger mechanism, naming each part.)

Q. What care must be taken of the magazines? A. The magazines must be given the best of care and kept in perfect condition. They will be kept cleaned and well oiled. They will not be dented or bent.

Q. How are the magazines filled? A. By placing the magazine filler over the magazine and loading four clips in the same manner as in the service rifle.

Q. How are the magazines loaded into the automatic rifle? A. (Man being examined loads automatic rifle and releases magazine.)

Q. What do the letters *F, A,* and *S* on the receiver mean? A. The letter *F* means semiautomatic fire; the letter *A* means automatic fire; and the letter *S* means safety.

Q. Why should the accessory and spare-parts kit be carried? A. Because it contains equipment which is necessary to keep the automatic rifle in action.

Q. How is the firing pin removed without disassembling the automatic rifle? A. (Man being examined shows how it is done.)

Q. How is the extractor removed without disassembling the automatic rifle? A. (Man being examined shows how it is done.)

Q. Why does the automatic rifle fire a single shot when the change lever is turned to *F*? A. Because the lug on the shank of the change lever is turned away from the toe of the connector, allowing the connector to rise until the cam surface on the rear of the head of the connector strikes the cam surface on the sear carrier, thereby forcing the head of the connector out from under the sear.

Q. Why cannot the trigger be pulled when the change lever is set on *S*? A. Because the change lever is turned so that the solid portion of the shank of the change lever is struck by the shoulders of the trigger.

Q. What is the first thing to do in case of any stoppage? A. Tap the magazine, pull back and push forward the operating handle, aim, and try to fire again.

Q. What is the next thing to do in case you have tried to fire again and the stoppage recurs? A. Pull back the operating handle slowly until it strikes the hammer pin and see what the position of the stoppage is; drop out the magazine; then apply immediate action.

Q. What are the three stoppages in the first position? A. Failure to feed, failure to fire, and insufficient gas.

Q. What things may cause failure to fire in the first position? A. Defective ammunition, broken firing pin, weak recoil spring, and too much friction.

Q. What usually causes failure to feed? A. Magazine trouble.

Q. How can you tell when the automatic rifle is not getting enough gas? A. The rifle will fire but the bolt will not go to the rear.

Q. If your automatic rifle is giving you trouble, due to failure to extract, what is probably the trouble? A. The chamber of the rifle has not been properly cleaned.

Q. In what way does cleaning the automatic rifle after firing differ from cleaning the service rifle? A. It is necessary to clean the piston, gas, cylinder, chamber, and magazines, as well as the bore.

Q. What solution is used to remove powder and primer mixture fouling from barrel?   *A.* Soda ash solution or hot soapy water.

Q. How should the automatic rifle be oiled?   *A.* The automatic rifle should be oiled by wiping a thin film of approved oil over all parts after cleaning.   When firing is done in the field, a light film of oil should be placed on the working parts, especially points where a great deal of friction occurs.

Q. Why cannot the barrel be removed and cleaned from the breech?   *A.* It is very difficult to replace the barrel as tightly as it should be and as soon as the barrel works loose the rifle will develop head space trouble.   When replacement is necessary it should be turned in.

c. Preparatory exercises.—The examination on section II, chapter 2, will consist of questions and demonstrations designed to test the soldier's knowledge and proficiency in the preparatory exercises.

d. The form shown below will be completed, for each individual who is to fire, as an essential part of this examination.

Form showing state of instruction

Names	Disassembling and assembling	Care and cleaning	Knowledge of functioning	Operation of the piece	Immediate action and stoppages	Sighting bar (1st sighting and aiming exercise)	Alining sights (2d sighting and aiming exercise)	Making shot groups (3d sighting and aiming exercise)	Prone position (with and without muzzle rest)	Sitting position	Kneeling position	Assault fire position	Trigger squeeze exercise	Rapid-fire exercise	Taking windage	Taking elevation	Understanding of the zero of a rifle	Use of the score book	Final grade
Brown																			
Jones																			
Smith																			
Knox																			

99

Fair	Good	Very good	Excellent	Excellent and has instructional ability
X	X X	X X X	X X X X	X X X X X

SECTION III

COURSES TO BE FIRED

■ 76. SCOPE AND OBJECT OF RANGE PRACTICE.—*a.* Range practice is divided into two phases:

(1) 1,000-inch range practice.

(2) Known-distance range practice.

b. Practice on the 1,000-inch range is included in all marksmanship courses to conserve time, ammunition, and troop labor during the range season. The 1,000-inch range provides a convenient short-distance range whereon the soldier can receive training with service ammunition in the fundamentals of automatic rifle marksmanship. Firing on the 1,000-inch range will be included in instruction practice for every individual firing a qualification course. The amount of such firing within the limits set forth in the tables will be determined by the company or higher commander. In general, recruits will require more of this type of firing in their instruction than previously qualified men. *All range firing will be semiautomatic fire.* The sandbag will be used for the muzzle rest prescribed in the firing tables.

■ 77. SEQUENCE OF FIRING.—The instruction practice outlined for each course is intended to serve as a guide. Variations may be made in the sequence prescribed within instruction practice to take advantage of time, weather, and range facilities. Variations may be made in the sequence prescribed within record practice for the same reasons. In no case will an individual's record practice in a course be interspersed with his instruction practice.

■ 78. MARKSMANSHIP COURSES.—One of the following courses will be fired by each automatic rifleman. The conduct and rules governing these courses are covered in sections IV and V

100

of chapter 2. The particular course to be fired will be designated by higher authority in accordance with the provisions of AR 775-10.

  a. Course A.—(1) *1,000-inch range.*—(a) *To zero rifle.*

TABLE I

Range (inches)	Time	Shots	Target	Position	Remarks
1,000...	No limit.	5	No. 1, 1,000-inch target.	Prone, muzzle rest.	
1,000...	...do.......	5	No. 2, 1,000-inch target.	Prone..........	Loop sling.
1,000...	...do.......	5	No. 3, 1,000-inch target.	Kneeling........	Hasty sling.

  (b) *Instruction practice.*

TABLE II

Range (inches)	Time (seconds)	Shots	Target	Position	Remarks
1,000...	15.........	5	No. 1, 1,000-inch target.	Prone, muzzle rest.	
1,000...	15.........	5	No. 2, 1,000-inch target.	Sitting.........	Hasty sling.
1,000...	15.........	5	No. 3, 1,000-inch target.	Kneeling.......	Do.
1,000...	55.........	12	Nos. 1, 2, 3, and 4, 1,000-inch target.	Prone, muzzle rest.	Four magazines of three rounds each.

  (2) *Known-distance range.*—(a) *Instruction Practice.*

TABLE III

Range (yards)	Time	Shots	Target	Position	Remarks
200	No limit..	5	Rifle D.....	Prone, muzzle rest...	
200	....do.....	5	.....do.......	Kneeling...............	Hasty sling.
300	....do.....	5	.....do.......	Prone, muzzle rest...	
500	....do.....	5	.....do.......	Do.	

TABLE IV

Range (yards)	Time (seconds)	Shots	Target	Position	Remarks
200	No limit	5	Rifle D	Prone	Loop sling.
200	do	5	do	Kneeling	Hasty sling.
300	do	5	do	Prone	Loop sling.
300	12	5	do	Prone, muzzle rest	
500	No limit	5	do	Prone	Loop sling.
500	15	5	do	Prone, muzzle rest	

TABLE V

Range (yards)	Time (seconds)	Shots	Target	Position	Remarks
200	No limit	5	Rifle D	Kneeling	Hasty sling.
200	30	9	do	Prone, muzzle rest.	Three magazines of three rounds each.
300	No limit	5	Rifle D	Prone	Loop sling.
300	35	9	do	Prone, muzzle rest.	Three magazines of three rounds each.
500	No limit	5	do	Prone	Loop sling.
500	40	9	do	Prone, muzzle rest.	Three magazines of three rounds each.

TABLE VI

Range (yards)	Time (seconds)	Shots	Target	Position	Remarks
200	35	15	Rifle D	Prone, muzzle rest.	Three magazines of five rounds each.
300	40	15	do	do	Do.
500	45	15	do	do	Do.

102

(b) *Record practice.*

TABLE VII

Range (yards)	Time (seconds)	Shots	Target	Position	Remarks
200	No limit	5	Rifle D	Kneeling	Hasty sling.
200	35	15	do	Prone, muzzle rest.	Three magazines of five rounds each.
300	No limit	5	do	Prone	Loop sling.
300	40	15	do	Prone, muzzle rest.	Three magazines of five rounds each.
500	No limit	(1)	do	Prone	Loop sling.
500	45	15	do	Prone, muzzle rest.	Three magazines of five rounds each.

1 2 sighting shots, 5 for record.

b. *Course B.*—(1) *1,000-inch range.*— Fire tables I and II of course A.

(2) *Known-distance range.*—(a) *Instruction practice.*

TABLE VIII

Range (yards)	Time	Shots	Target	Position	Remarks
200	No limit	5	Rifle D	Prone	Loop sling.
200	do	5	do	Kneeling	Hasty sling.
300	do	5	do	Prone	Loop sling.

TABLE IX

Range (yards)	Time (seconds)	Shots	Target	Position	Remarks
200	No limit	5	Rifle D	Prone, muzzle rest.	
200	do	5	do	Kneeling	Hasty sling.
300	do	5	do	Prone	Loop sling.
300	12	5	do	Prone, muzzle rest.	

103

TABLE X

Range (yards)	Time (seconds)	Shots	Target	Position	Remarks
200	No limit	5	Rifle D	Kneeling	Hasty sling.
200	30	9	....do	Prone, muzzle rest.	Three magazines of three rounds each.
300	No limit	5	....do	Prone	Loop sling.
300	35	9	....do	Prone, muzzle rest.	Three magazines of three rounds each.

TABLE XI

Range (yards)	Time (seconds)	Shots	Target	Position	Remarks
200	35	15	Rifle D	Prone, muzzle rest.	Three magazines of five rounds each.
300	40	15	....do	....do	Do.

(b) *Record practice.*

TABLE XII

Range (yards)	Time (seconds)	Shots	Target	Position	Remarks
200	No limit	5	Rifle D	Kneeling	Hasty sling.
200	35	15	....do	Prone, muzzle rest.	Three magazines of five rounds each.
300	No limit	5	....do	Prone	Loop sling.
300	40	15	....do	Prone, muzzle rest.	Three magazines of five rounds each.

c. *Course C.*—(1) *1,000-inch range.*—Fire tables I and II of course A.

(2) *Known-distance range.—(a) Instruction practice.*

TABLE XIII

Range (yards)	Time	Shots	Target	Position	Remarks
200	No limit.	5	Rifle D_____	Prone_____	Loop sling.
200	___do_____	5	_____do_____	Kneeling_____	Hasty sling.

TABLE XIV (fire twice)

Range (yards)	Time (seconds)	Shots	Target	Position	Remarks
200	No limit.	5	Rifle D_____	Prone_____	Loop sling.
200	___do_____	5	_____do_____	Kneeling_____	Hasty sling.
200	35_____	15	_____do_____	Prone, muzzle rest.	Three magazines of five rounds each.

(b) *Record practice.*

TABLE XV

Range (yards)	Time (seconds)	Shots	Target	Position	Remarks
200	No limit.	5	Rifle D_____	Prone_____	Loop sling.
200	___do_____	5	_____do_____	Kneeling_____	Hasty sling.
200	50_____	30	_____do_____	Prone, muzzle rest.	Three magazines of ten rounds each;

d. Course D—1,000-inch range.—(1) Instruction practice.—
Fire tables I and II of course A.

(2) *Record practice.*—Fire table II of course A.

SECTION IV

RANGE PRACTICE

■ 79. GENERAL.—*a. Training programs and schedules.*—Training programs and schedules will provide a period for range practice.

105

b. Range practice.—Range practice includes both 1,000-inch firing and known-distance firing.

c. Officers' range practice.—The officers of an organization will be enabled to complete their own range practice in advance of their men whenever practicable in order that their entire attention may be given to their instructional duties.

d. Uniform.—The uniform to be worn during instruction practice and record practice will be prescribed by the commanding officer. The automatic rifleman's belt will be worn during instruction practice and record practice.

e. Use of pads.—The use of elbow pads is recommended. The use of shoulder pads is unnecessary but is permitted.

■ 80. ORGANIZATION.—*a. Officer in charge of firing.*—An officer in charge of firing will be designated by the responsible commander. The officer in charge of firing, or his deputy, will be present during all firing and will be in charge of the practice and safety precautions on the range.

b. The officers, noncommissioned officers, and coaches of the units on the range will perform duties generally similar to those prescribed for them in preparatory marksmanship instruction. See section II.

c. Range officer.—A range officer with such commissioned and enlisted assistants as are necessary will be appointed by the post or station commander well in advance of range practice. At large camps or stations where the coordination of range practice for different organizations is involved, he may function as the direct representative of the camp or station commander. In other cases he should be made responsible to the officer in charge of firing and in all cases should cooperate closely with him. The range officer will make timely estimates for material and labor to place the range in proper condition for range practice, and will supervise and direct all necessary repairs to shelters, butts, targets, firing points, and telephone lines. He exercises direct supervision over the practical operation of the rifle range during the practice season. He regulates the distribution of ranges and targets and, in general, assists the officer in charge of firing by using the means necessary to secure efficient and accurate service from the working force of the range. He provides safe conditions for the markers and any visitors. Whenever necessary, he provides range guards and instructs them in the methods to be used for the protection of life and property in the danger area.

d. Unit range officer.—During the operation of any range by a unit the commanding officer thereof may detail an officer as unit range officer. The unit range officer will be responsible to the commanding officer of the organization to which the range is assigned for its efficient operation.

e. Pit detail.—An officer or noncommissioned officer with such assistants as may be necessary will be detailed in charge of arrangements in the pit. He will be responsible to the officer in charge of firing for the discipline, efficiency, and safety of all pit details. He sees that all of the target equipment is kept in serviceable condition; that the desired targets are ready for firing at the appointed time, and that all target details are provided with the proper flags, marking disks, paste, pasters, and spotters.

f. Use of dummy cartridges.—The corrugated type of dummy cartridges may be used in range practice. When ammunition must be conserved, a proportion of the corrugated type dummies may be included in magazines with live ammunition. The use of any other type of range dummies is prohibited.

■ 81. USE OF SANDBAG REST.—The sandbag rest is used in all rapid firing in the prone position.

■ 82. FIRING POINTS.—All firing points should have firm soil. Loose loam or sand on the firing point has an adverse effect on accuracy.

■ 83. COACHING.—*a. General.*—(1) During instruction practice each man on the firing line will have a coach to watch him and to help him correct his errors. An average soldier who has been properly instructed in the preparatory work or who has been given instruction in coaching methods can be used with good results and should be used when more experienced coaches are not available.

(2) It is a good practice to have experienced coaches in charge of one or more targets, usually on a flank, to which pupils are sent for special coaching if required.

(3) Great patience will be exercised by the coach so as not to excite or confuse the firer. Everything is done to encourage him. It is often a good plan to change coaches. It is necessary to do so if the coach shows signs of impatience.

b. Position of coach.—The coach will take the same position as the man who is firing; that is, prone, sitting, or kneeling. This enables the coach to watch the pupil's trigger finger and his eye.

c. Duties of coach.—The success of the instruction will depend to a great extent on the thoroughness and exactness with which the coach performs his duties. During firing the duties of the coach in addition to those given in paragraph 57 are as follows:

(1) To require the firer to inspect his rifle.

(2) To check the sight setting and aiming, requiring them to be correct.

(3) To observe the firer and see that he re-aims after each shot.

(4) To require the firer to fire as required for each target.

(5) To point out errors and explain their effect upon the exercises.

(6) To keep constant watch on the adjustment and condition of the gun.

■ 84. To ALINE FRONT SIGHT.—*a.* As the Browning automatic rifle, caliber .30, M1918, is not equipped with a wind gage on the rear sight, provision has been made whereby the front sight may be tapped to the right or left so that it will not be necessary to aim off on account of a defective alinement of the sights. The 1,000-inch range is the best place to do this because the aiming point is small and well defined and atmospheric conditions will have no effect on the flight of the bullet.

b. In firing, the soldier aims at the bottom edge of a designated figure on the 1,000-inch target, U. S. rifle, caliber .30, M1. He fires three shots semiautomatically and very carefully, using exactly the same aiming point for all three. The center of the resulting shot group indicates how much and in what direction to move the front sight. The soldier then verifies the front sight adjustment by firing two more rounds.

c. If the shot group is to the right of the aiming point, move the front sight to the right; if the left, move the front sight to the left. It should be borne in mind that to move the strike of the bullet 1 inch at a range of 1,000 inches requires a movement of the front sight of only twenty-five one-thousandths of an inch.

d. When the same rifle is being used by a group, the alinement of the front sight will be executed by any soldier in the group who has qualified in the record course with the automatic rifle. If the group contains no such man, the firing and alinement of the sight will be executed by any member of the group holding the highest qualification with the service rifle.

e. The front sight of the rifle will not be moved after it is once zeroed, except by authority of an officer.

■ 85. To Determine Zero of Rifle.—*a.* After the front sight has been alined as prescribed in paragraph 84, to zero the automatic rifle on the 1,000-inch target, a correct and steady aim must be taken on a designated figure with the rear sight set at an elevation of 300 yards for the first shot. Correction in elevation to place the shots in the center of the scoring space may be applied to the rear sight under the direction of the coach after each shot.

b. For the zeroing on the known-distance range, the targets should be pulled after each shot of semiautomatic fire and a spotter placed in the target to facilitate the correction of errors in the sights.

c. When the same automatic rifle is being used by a group, the highest qualified man with the service rifle in the group will zero the weapon as far as the movement of the front sight is concerned. Each member of the group should use his zeroing ammunition to determine his zero elevation, and the exact point of aim necessary for him to place his shots in the center of the scoring space.

■ 86. Instruction Practice on 1,000-Inch Range.—*a.* Instruction practice on the 1,000-inch range will conform to the regulations given in section V of this chapter for record practice except that coaching is permitted and additional personnel to score targets are not required.

b. Each exercise on the 1,000-inch range will be preceded by an appropriate fire order.

c. Form of fire order for slow fire, 1,000-inch range.

(1) Announce the position.

(2) Slow fire, figure 1.

(3) _____ rounds, with ball ammunition, LOAD.

(4) COMMENCE FIRING.

(5) CEASE FIRING.

(6) CLEAR RIFLE.

d. Form of fire order for rapid fire, 1,000-inch range.

(1) Announce the position, number of magazines, and number of rounds per magazine.

(2) With ball cartridges, LOAD.

(3) Rapid fire, figures 1 and 2. _____ rounds on each figure .

(4) COMMENCE FIRING.

(5) CEASE FIRING.

(6) CLEAR RIFLE.

e. If the 1,000-inch range is equipped with pits and sliding targets similar to those of a known-distance range, the fire orders for rapid fire prescribed for the known-distance range under paragraph 87*d* apply except that *d*(3) above is substituted for paragraph 87*d*(3).

■ 87. INSTRUCTION PRACTICE ON KNOWN-DISTANCE RANGE.—*a.* Instruction practice is carried out in conformity with the regulations governing record practice as given in section V, except that additional personnel for scoring targets are not required and each firer will have a coach with him on the firing line. The officer in charge of firing may prescribe the sequence of firing the courses of instruction practice.

b. Each exercise on the known-distance range will be preceded by an appropriate fire order.

c. Form of fire order for slow fire, known-distance range.

(1) Announce the position and number of rounds to be fired.

(2) With ball cartridges, LOAD.

(3) Slow fire.

(4) COMMENCE FIRING.

(5) CEASE FIRING.

(6) CLEAR RIFLE.

d. Form of fire order for rapid fire, known-distance range.

(1) Announce the position, number of magazines, and number of rounds per magazine.

(2) With ball cartridge, LOAD.

(3) Rapid fire.

(4) Ready on the right?

(5) Ready on the left?

(6) READY ON THE FIRING LINE?

(7) CEASE FIRING.

(8) CLEAR RIFLE.

The targets are withdrawn before the exercise starts and the red flag displayed at the center target. The command READY ON THE FIRING LINE is transmitted to the officer or noncommissioned officer in charge in the pits who will have the red flag waved and lowered on its receipt and who will cause the targets to be run up simultaneously 5 seconds after the flag is lowered. Upon the expiration of the proper time interval he causes the targets to be withdrawn. The officer in charge of the firing line gives the commands CEASE FIRING and CLEAR RIFLE when targets are withdrawn.

■ 88. SAFETY PRECAUTIONS.—Safety precautions for observance by troops are self-contained and complete in this manual. Reference to AR 750–10 is necessary for range officers, the officer in charge of firing, and the commander responsible for the location of ranges and the conduct of firing thereon. All officers and men who are to fire or who are concerned with range practice will be familiarized with the following safety precautions before firing is commenced:

a. Danger flags will be displayed at prominent positions on the range during firing. Do not fire unless such flags are displayed.

b. Upon arrival at the range the automatic rifles of an organization will be inspected by the officers to see that chambers and barrels are free from obstruction.

c. All rifles on the range except those in use on the firing line will have bolts in the forward position and magazines withdrawn. Rifles on the firing line will not be loaded without command.

d. Consider every rifle to be loaded until it is examined and found to be unloaded. Never trust your memory as to its condition in this respect.

e. Never point the rifle in any direction where an accidental discharge may cause harm while the bolt is in its rear position.

f. Firing will not begin on any range until the officer in charge of firing has ascertained that the range is clear, and has given the commands LOAD and COMMENCE FIRING.

g. At least one officer will be present at all firing.

h. No rifle will be removed from the firing line until an officer has inspected it to see that the bolt is in its forward position and the magazine is withdrawn.

i. No person will be allowed in front of the firing line for any purpose until directed by an officer who has ordered all rifles to be cleared and ascertained that the order has been carried out.

j. All firing will immediately cease and rifles set at "safe" (or cleared if ordered) at the command CEASE FIRING.

k. All loading and unloading will be executed on the firing line with the muzzle directed toward the targets. Rifles will never be loaded in rear of the firing line.

l. Care should be taken to avoid undue exposure of ammunition to the direct rays of the sun. This creates hazardous chamber pressures.

m. Never grease or oil the ammunition or the walls of the rifle chamber.

n. See that the ammunition is clean and dry. Examine all live *and dummy ammunition.* Turn in all cartridges with loose bullets or which appear to be otherwise defective.

o. Never fire a rifle with any rust-preventive compound, cleaning patch, dust, dirt, mud, snow, or other obstruction in the bore. To do so may burst the barrel.

p. Before leaving the range all rifles and belts will be inspected by an officer to see that they do not contain ammunition, and men in ranks will be questioned as to whether they have any ammunition in their possession.

q. See AR 45–30 for regulations covering *report of accident* involving ordnance matériel.

r. No magazine test or magazine drill will be conducted in rear of the firing line.

SECTION V

REGULATIONS GOVERNING RECORD PRACTICE

■ 89. GENERAL.—*a*. Record practice for course A, B, or C is fired on the known-distance range. Record practice for course D is fired on the 1,000-inch range. Additional provisions applicable for course D are given in paragraph 119.

b. Record practice will follow instruction practice.

c. When the record practice of an individual has commenced it will be completed without interruption by any other form of firing. Instruction practice and record practice will not be conducted simultaneously unless the two types of practice are conducted on different parts of the range.

d. The officer in charge of firing may, at his discretion, require record practice upon the day on which instruction practice is completed.

■ 90. FIRE ORDERS.—Every rapid-fire exercise fired in record practice will be preceded by an appropriate fire order. Suitable forms for such orders are given in section IV of this chapter.

■ 91. SEQUENCE OF EXERCISES.—The exercises given in the table for record practice will be fired in the sequence directed by the officer in charge of firing.

■ 92. STOPPAGES.—*a*. When a stoppage occurs which cannot be cleared by operating the operating handle, the firer will call, "Time." The officer in charge of firing, or one of his assistants, will note the time left to complete the exercise and investigate the stoppage. The stoppage will be reduced. If the stoppage was not due to any fault of the firer, he will be authorized to load, aim, and commence firing on command from the officer investigating the stoppage who will allow him the unexpired time. In cases where the exact time remaining was not determined by the officer in charge, the firer will be allowed 2 seconds per round for the remaining rounds. When time and ammunition permit, the complete exercise will be refired.

b. If the stoppage is manifestly the fault of the firer in failing to inspect either the gun, magazines, or ammunition, or due to incorrect loading, or replacing of magazines, no time

113

will be allowed to complete the firing and only that part of the exercise which was completed will be scored.

c. The firer will be allowed to fire rounds ejected in clearing stoppages.

d. The soldier firing must not be given any information with reference to the location of his previous hits on the incompleted target until the score is completed.

e. Should a breakage occur, the gun will be repaired or a different gun substituted and the exercise completed. If a different rifle is substituted, the firer will be allowed extra rounds to determine the zero of the substituted rifle.

f. The officer in charge, or his assistants, will render all decisions on stoppages.

g. A firer, firing part of a rapid fire exercise, will begin his firing with the entire target exposed.

■ 93. MEN MARKING TARGETS NOT TO KNOW WHO IS FIRING.— Officers and men in the pit will not be informed as to who is firing on any particular target. In case of such violation the firer will be required to repeat his score and appropriate disciplinary action taken.

■ 94. TARGET DETAILS.—The details in the pit for the supervision, operation, marking, and scoring of targets during record practice consist of officers, noncommissioned officers, and privates as follows:

a. One commissioned officer assigned to each two targets. When it is impracticable to detail one officer to each two targets in the pit, an officer will be assigned to supervise the marking and scoring of not to exceed four targets. In this case the pit scores will be kept by the noncommissioned officer in charge of each target who will sign the score cards. The officer will take up and sign each score card as soon as the complete course is recorded.

b. One noncommissioned officer assigned to each target to direct and supervise the markers. This noncommissioned officer will be selected from an organization other than the one firing on the target which he supervises. If this is not possible, the officer assigned to the target will exercise special care to insure correct scoring.

c. One or two privates assigned to operate and mark each target. These privates may be selected from the organization firing on the targets to which they are assigned.

■ 95. ORGANIZATON OF FIRING LINE.—*a.* Scorers seated close to and to the right of the person firing.

b. Telephone operators, 5 paces in rear of the firing line.

c. Persons awaiting their turn to fire, 10 paces in rear of the firing line.

d. Low arm racks or rifle racks and cleaning racks, 20 paces in rear of the firing line.

■ 96. SCORE CARDS AND SCORING.—*a.* Duplicate score cards will be kept, one at the firing point and one in the pit. These cards will be of different colors. The cards at the firing point will bear the date, the man's name, the number of the target, and the order of firing. The pit record card will not show the man's name, but will bear the date, the number of the target, and the order of firing.

b. Entries on all score cards will be made in ink or with indelible pencil. No alterations or corrections will be made on the card except by the organization commander or officer scorer in the pit who will initial each such alteration or erasure made.

c. The scores at each firing point will be kept by a non-commissioned officer of some organization other than the organization firing on the target to which he is assigned. If this is not possible company officers will exercise special care to insure correct scoring. As soon as the score is completed the score card will be signed by the scorer, taken up, and signed by the officer supervising the scoring. When convenient the score cards are turned over to the organization commander. Except when required for entering new scores on the range, score cards will be retained in the personal possession of the organization commander.

d. In the pit the officer keeps the scores for the targets to which he is assigned. As soon as the score is completed he signs the score card. He turns these cards over to the organization commander at the end of the day's firing. The organization commander will check the pit records against the firing-line records. In case of discrepancy between the two the pit record governs.

e. Upon completion of the record firing and after the qualification order is issued, the pit score card of each man will be attached to his score card kept at the firing point. These cards will be kept available for inspection among the company records for 1 year and then destroyed.

■ 97. Marking.—*a. Slow fire.*—(1) The value of the shot is indicated as follows:

(*a*) A five by a white disk.

(*b*) A four by a red disk.

(*c*) A three by a white disk with a black cross.

(*d*) A two by a black disk.

(*e*) A miss by waving a red flag across the front of the target.

(*f*) Ricochet hits will be counted as a miss and so indicated.

(2) The exact location of the hit is indicated by placing a spotter of appropriate size in the shot hole. The center of the marking disk is placed over the spotter in marking hits.

b. Rapid fire.—(1) The disks described in *a* above are used to indicate the value of the hits.

(2) Spotters are placed in shot holes before running the target up for marking.

(3) The marking begins with the hits of the highest value. The center of the disk is placed over the spotter, then swung off the target and back again to the next spotter, care being taken each time to show the correct face of the disk. The marking must be slow enough to avoid confusing the scorer at the firing point. When one spotter covers more than one shot hole the disk is placed over it the required number of times. Misses are indicated by slowly waving the red flag across the face of the target, one time for each miss.

■ 98. Procedure in Slow Fire.—*a. On firing line.*—(1) One person will be assigned to each target in each order.

(2) As the value of each shot is signaled, the scorer announces the following data in a tone sufficiently loud to be heard by the firer: The name of the firer, the number of the shot, and the value of the hit. The scorer then records the value of the hit on the score card.

(3) Whenever a target is marked before the individual who is assigned thereto has fired, as will occur when another man fires on the wrong target, the scorer will notify the officer in charge of firing. The latter will notify the officer in the pit assigned to the target to disregard that shot. This precaution is necessary to prevent errors in the pit record.

(4) When an individual fires on the wrong target he will not be assigned a miss until the target to which he is assigned has been pulled down and the miss signaled from the pit.

(5) If the target is not half-masked at the completion of a score thereon, or if it is half-masked at the wrong time, the officer in charge of that firing point will adjust the matter at once over the telephone. This precaution is necessary to prevent the error from being carried on through the scores that follow.

b. *In pit.*—(1) The target is withdrawn and marked after each shot.

(2) When a shot is fired at a target, it is pulled down. The noncommissioned officer indicates the location of the hit to the officer assigned to the target who announces its value and records it on the score card. A spotter is then placed in the shot hole, the previous shot hole, if any, is pasted, and the target is run up and marked. The noncommissioned officer supervises the marking of each shot. The officer also exercises general supervision over the marking.

(3) When the pit score card indicates a score has been completed, the target is half-masked for about 30 seconds as a signal to the firing line of such completion. At the end of the 30 seconds the target is pulled fully down, the spotter removed, the shot hole pasted, and the target run up for the beginning of a new score.

(4) When a target frame is used as a counterweight for a double sliding target, the blank side of such frame will be toward the firing line.

■ 99. PROCEDURE FOR RAPID FIRE.—*a. On firing line.*—(1) One person only will be assigned to a target in each order.

(2) When all is ready in the pit the red flag is displayed at the center target. At that signal the officer in charge

of the firing line will conduct the exercise to be fired in accordance with the procedure given in section IV.

(3) If any individual fails to fire at all, he will be given another opportunity. If he fires one or more shots, the score must stand as his record except as provided in paragraph 92. He will not be permitted to repeat his score on the claim that he was not ready to fire.

(4) As each shot is signaled from the pits, it is announced by the scorer at the firing line. A score of 15 shots is announced as follows as each shot is marked: "Target 22; 1 five, 2 fives, 3 fives, 4 fives, 5 fives, 6 fives, 7 fives; 1 four, 2 fours, 3 fours, 4 fours, 5 fours, 6 fours, 7 fours; 1 two." The scorer notes these values on a pad and watches the target as he calls the shot. After marking is finished he counts the number of shots marked, and if it is more or less than 15 calls, "Re-mark No. _____." If 15 shots have been marked, he then enters the value of each hit and their total value on the soldier's score card.

b. In pit.—(1) The time of fire allowed for each exercise is regulated by the officer in charge of the pit. The procedure is as follows: The targets are withdrawn before the exercise starts, and the red flag is displayed at the center target. The command READY ON THE FIRING LINE is transmitted to the officer or noncommissioned officer in charge in the pits who will have the red flag waved and lowered on its receipt and who will cause the targets to be run up simultaneously 5 seconds after the flag is lowered. Upon the expiration of the proper time interval, he causes the targets to be withdrawn.

(2) The officers scoring in the pit examine each of their targets in turn, announce the score, and record it on the pit score cards. Spotters are then placed in the shot holes and the targets run up and marked. The noncommissioned officers supervise the marking of each shot. The officers exercise general supervision over the marking of their targets.

(3) The targets are left up for about 1 minute after being marked. They are then withdrawn, pasted, and made ready for another score. They may be left up until ordered pasted by the officer in charge of the firing line.

(4) If more than 15 shots are found on any target in record practice, it will not be marked unless all of the hits are of the same value. The officer in charge of the firing line will be notified of the facts by telephone.

■ 100. USE OF TELEPHONES.—*a.* Telephones will be used for official communication only.

b. No one will ask over the telephone for information as to the name or organization of any person firing on any particular target, and no information of this nature will be transmitted.

c. The following expressions will be used over the telephone in the cases enumerated:

(1) When a shot has been fired and the target has not been withdrawn from the firing position, "Mark No. _ _ _ _ _ _."

(2) When a shot has been fired and a target has been withdrawn from the firing position but not marked, "Disk No. _ _ _ _ _ _."

(3) When the target has been withdrawn from the firing position and marked but the value of the shot has not been understood, "Re-disk No. _ _ _ _ _ _."

(4) When the firing line is ready for rapid fire, "Ready on the firing line."

(5) When a shot is marked on a target and the person assigned thereto has not fired, "Disregard the last shot on No. _ _ _ _ _ _."

■ 101. COACHING PROHIBITED.—Coaching of any nature after the firer takes his place on the firing line is prohibited. No person will render or attempt to render the firer any assistance whatever while he is taking his position or after he has taken his position at the firing line.

■ 102. USE OF INSTRUMENTS.—*a.* The use of field glasses is authorized and encouraged.

b. The use of instruments or devices for determining the force and direction of the wind is prohibited during record practice.

■ 103. SHELTER FOR FIRER.—Sheds or shelter for the individual at the firing point will not be permitted at any range.

■ 104. RESTRICTIONS AS TO RIFLE.—Troops will use the Browning automatic rifle, caliber .30, M1918, as it is issued by the Ordnance Department. The use of additional appliances is prohibited. The sights may be blackened. Ordnance Department test equipment will not be used for determining the classification.

■ 105. AMMUNITION.—The ammunition used will be the service cartridge as issued by the Ordnance Department.

■ 106. CLEANING.—Cleaning is permitted at any time.

■ 107. USE OF GUN SLING.—The gun sling will be used as prescribed for the various positions in these regulations, and in no other manner.

■ 108. PADS AND GLOVES.—a. Pads of moderate size and thickness may be worn on both elbows to protect them from bruising. A smooth pad of moderate size and thickness may be worn on the right shoulder. The use of other forms of pads is prohibited. The use of a hook or ridge on the sleeve of the shooting coat or shirt to keep the sling in place on the arm is prohibited.

b. A glove may be worn on either hand provided it is not used to form an artificial support for the rifle.

■ 109. WARMING, FOULING, AND SIGHTING SHOTS.—No warming or fouling shots will be allowed. Two sighting shots are authorized and required at 500 yards, slow fire.

■ 110. SHOTS CUTTING EDGE OF SILHOUETTE OR LINE.—Any shot cutting the edge of the silhouette will be indicated and recorded as a hit in the silhouette. Because the limiting line of each division of the target is the outer edge of the line separating it from the exterior division, a shot touching this line will be indicated and recorded as a hit in the higher division.

■ 111. SLOW FIRE SCORE INTERRUPTED.—If a slow fire score is interrupted through no fault of the person firing, the unfired shots necessary to complete the score will be fired at the first opportunity.

■ 112. MISSES.—Before a miss is signaled in record firing the target will be withdrawn and carefully examined by an officer. Whenever a target is run up and a miss is indicated, it will be presumed that this examination has been thoroughly made. No challenge of the value indicated will be entertained, or resignaling of the shot allowed.

■ 113. SHOTS TO BE INCLUDED IN SCORE.—All shots fired by the soldier in his proper turn after he has taken his place at the firing line and the target is ready will be considered as part of his score.

■ 114. FIRING ON WRONG TARGET.—Shots fired on the wrong target will be recorded as a miss on the score of the man firing, no matter what the value of the hit on the wrong target may be. In rapid fire the soldier at fault is credited with only such hits as he may have made on his own target.

■ 115. TWO SHOTS ON SAME TARGET.—In slow fire, if two shots strike a target at the same time, or nearly the same time, neither will be marked. The individual who fired on his own target will be allowed another shot.

■ 116. WITHDRAWING TARGET PREMATURELY.—In slow fire, if the target is withdrawn from the firing position just as a shot is fired, the scorer at that firing point will at once report the fact to the officer in charge of the scoring on that target. The officer will investigate to see if the case is as represented. Being satisfied that such is the case, he will direct the shot be disregarded and that the man fire another shot.

■ 117. UNFIRED CARTRIDGES IN RAPID FIRE.—Each unfired cartridge will be recorded as a miss.

■ 118. MORE SHOTS THAN PRESCRIBED IN RAPID FIRE.—When a target has more than the prescribed number of shots for a rapid-fire exercise in record practice and these hits are of different values the target will not be marked. The soldier firing on that target will repeat the firing of his score. If all the hits on the target have the same value the target will be marked and he will be given the value of the authorized number of shots.

■ 119. RECORD PRACTICE FOR COURSE D, 1,000-INCH RANGE.—a. The following special provisions apply only to record practice for course D, which is fired on the 1,000-inch range.

121

b. So much of the foregoing regulations for record practice as can be applied will be followed. Suitable fire orders for use on the type of 1,000-inch range which is equipped with pits and movable targets, as well as suitable fire orders for use on the type of 1,000-inch range which is not so equipped, will be found in section IV.

c. When the record practice is fired on 1,000-inch ranges not equipped with pits and movable targets, the following rules will apply:

(1) Sufficient assistants will be detailed from companies other than the ones firing to assist the officer in charge. From the assistants, officers will be detailed as scorers at the rate of one for every four targets.

(2) The officers detailed as assistants will aid the officer in charge in every way possible. They will—

(*a*) Note deductions for penalties and report same to the scorer. (See *e*(3) below.)

(*b*) Note the time out for stoppages and inspect to determine whether the stoppage was due to any fault of the soldier.

(*c*) Superintend the firing of rounds remaining from stoppages not the fault of the firer.

(*d*) Scorers will count the bullet holes in each target and report any that have more than the prescribed number.

(*e*) Scorers will score the targets in accordance with the provisions of *e* below.

d. (1) When a stoppage occurs that cannot be cleared by pulling back the operating handle and releasing it, the firer will call, "Time." The officer in charge of firing, or an assistant, will note the time left to complete the exercise. The stoppage will be reduced. The firer will load and complete the firing on command from the officer in charge who will allow the remaining time. In cases where the exact time remaining was not determined by the officer in charge, the firer will be allowed 2 seconds per round for the remaining rounds.

(2) If the stoppage is manifestly the fault of the firer, no time will be allowed to complete the exercise and only that part of the exercise which was fired will be scored.

(3) Should a breakage occur, the rifle will be repaired or a different rifle substituted. If a different rifle is substituted, the firer will then be allowed four extra rounds to determine the zero of the substituted rifle. He will then complete the exercise.

e. The 1,000-inch record target will be scored in accordance with the requirements for record firing as follows:

(1) Hits within or cutting the line of the inner silhouette will count 5 points. Hits within or cutting the line of the middle silhouette will count 4 points. Hits within or cutting the line on the outside silhouette will count 3 points.

(2) Hits cutting the line of two scoring figures will be counted so as to give the firer the higher score.

(3) For firing before COMMENCE FIRING or after CEASE FIRING, 5 points will be deducted for each round so fired.

(4) In case of hits on the wrong target, the firer who received the erroneous hits will refire his score. The firer who placed his hits on the wrong target will count only those upon his own and will not be permitted to refire the exercise.

SECTION VI

TARGETS AND RANGES

■ 120. TARGETS.—The designations and dimensions of the two types of target used for marksmanship courses for the Browning automatic rifle, caliber .30, M1918, are as follows:

a. The 1,000-inch target, U. S. rifle, caliber .30, M1 (see fig. 39).—This target is used for fire on the 1,000-inch range. The scoring figures numbered from 1 to 4, inclusive, are utilized in known-distance marksmanship (ch. 2). The figures 5 to 8 are suitable for instruction in the technique of fire (ch. 5). Each of these scoring figures is composed of three silhouettes. These silhouettes are reduced in scale to represent the appearance of target D on the known-distance range. The inner silhouette of the 1,000-inch target represents the silhouette of target D at 500 yards. The middle silhouette of the 1,000-inch target represents the four space of target D at 300 yards. The outer silhouette of the 1,000-inch target represents the four space of target D at 200 yards. Hits in the inner silhouettes of the 1,000-inch target count 5, in the middle silhouettes 4, and in the outer silhouettes 3.

123

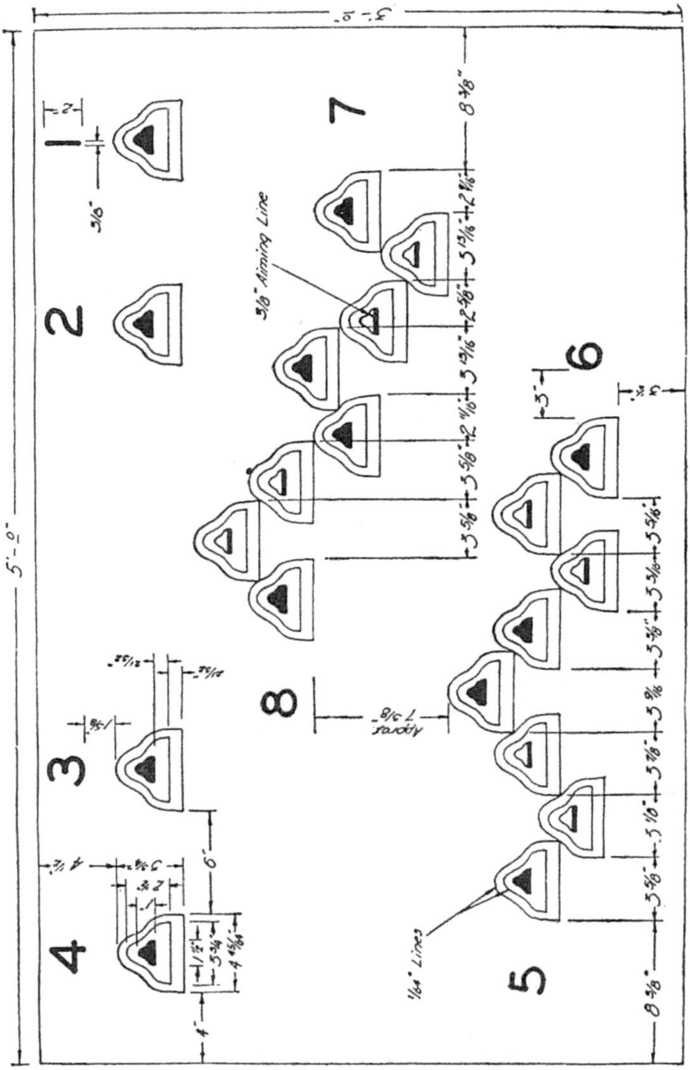

FIGURE 39.—Target, U. S. rifle, caliber .30, M1, 1,000-inch range

b. Target D.—This target is used for all slow and rapid fire on the known-distance range. It consists essentially of a square target, 6 by 6 feet in dimensions, upon which a black silhouette representing a prone figure is centered. Hits in the silhouette count 5, in the next space 4, and in the next 3. Hits on the remainder of the target count 2.

■ 121. RANGES.—*a. Suitability.*—Ranges suitable for range firing with the U. S. rifle, caliber .30, M1903 and M1, are equally suitable for range firing with the Browning automatic rifle, caliber .30, M1918.

b. Installation and construction.—The installation and construction of target ranges for small-arms target practice is governed by AR 30–1505. The installation of range communication systems is governed by AR 105–20. Range regulations for firing ammunition in time of peace are given in AR 750–10 and include the safety limits and danger areas of ranges. Information in regard to the selection of known-distance ranges is contained in *c* below.

c. (1) *Direction of range.*—If possible, a range should be so located that the firing is toward, or slightly to the east of north. Such location gives a good light on the face of the targets during the greater part of the day. However, security and suitable ground are more important than direction.

(2) *Best ground for range.*—The targets should be on the same level with the firer, or only slightly above him. Firing downhill should be avoided.

(3) *Size of range.*—The size of the range is determined by its plan and by the number of troops that will fire over it at a time. There are two general plans used in range construction; one with a single target pit and firing points for each range, the other with firing points on one continuous line, the target pits for the various ranges being in echelon.

(4) *Intervals between targets.*—The targets should be no farther apart than is necessary to reduce the chance of shots being fired on wrong targets. As a general rule, the intervals between targets are equal to the width of the targets themselves. Where the necessity exists for as many targets as possible in a limited space, this interval may be reduced one-half without materially affecting the value of the instruction.

(5) *Protection for markers.*—(a) Protection is provided for the pit details by excavating a pit, or by constructing a parapet in front of them, or by a combination of both methods.

(b) Where there are several targets in a row, the shelter should be continuous. It must be high enough to protect the markers. The parapet may be of earth with a timber or concrete revetment of sufficient thickness to stop bullets and from $7\frac{1}{2}$ to 8 feet high above the ground or platform on which the markers stand.

(6) *Artificial butts.*—If an artificial butt is constructed as a bullet stop, it should be of earth not less than 30 feet high and with a slope of not less than 45°. The slopes should be sodded. The provisions of AR 750–10 must also be met by the range.

(7) *Hills as butts.*—A natural hill to form an effective butt should have a slope of not less than 45°.

(8) *Numbering of targets.*—Each target should be designated by a number. The numbers for ranges up to 600 yards should be at least 6 feet in height and should be painted black on a white background. Arabic numerals of the size suggested will always be quickly recognized. They should be placed on the butt behind each target or on the parapet in front, and not so far above or below as to prevent the firer seeing the number when aiming at the target.

(9) *Measuring range.*—The range should be carefully measured and marked with stakes at the firing point in front of each target. These stakes should be about 12 inches above the ground and painted white. Black figures indicate the number of the corresponding target.

(10) *Ranges parallel.*—The different ranges for the same distance should all be parallel, so that similar conditions with respect to wind and light may exist.

(11) *Firing mounds.*—If it is necessary to raise a firing point, a low mound of earth no higher than required should be made. The mound should be level, sodded, and not less than 12 feet square. If the entire firing line is raised, the firing mound should be level, sodded, and not less than 12 feet wide on top.

126

(12) *Pit shed.*—A small house or shed should be built in or near the target pit for storing equipment.

(13) *Danger signals.*—A danger signal will be placed in front of the targets when firing has been suspended. One or more red streamers will be prominently displayed on all ranges and at all times during firing.

(14) *Range house.*—On large ranges a house containing a storeroom and office room is desirable.

(15) *Telephone service.*—Ranges should be equipped with a telephone system connecting the target pit with each firing point, the range house, and the station headquarters. The number of telephones should not be less than one to each ten targets.

(16) *Electric bells.*—On large ranges the installation for each five targets of an electric bell that can be controlled from a central point in the pit adds materially to the celerity and uniformity of target manipulation for rapid fire.

(17) *Covered ways between pits.*—Where the pits are in echelon, covered ways or tunnels should be provided between the various pits. This construction will allow the pit details to be shifted with safety without interrupting the firing.

(18) *1,000-inch range.*—A 1,000-inch range without a land danger area behind its backstop must meet the following minimum requirements.

(*a*) Vertical bullet-proof backstop and wing walls (natural or artificial) not less than 30 feet high. Wing walls must cover at least 15° on each flank. In case of artificial wing walls, they should be set at an angle of 15° with the backstop toward the firing points.

(*b*) Ricochet pit in front of firing points providing at least a 4° slope downward from the normal line of fire from a prone position and extending to within 30 feet of the backstop and wing walls. If a vertical cliff or wall over 40 feet high is available, no ricochet pit need be provided.

CHAPTER 3

MARKSMANSHIP—MOVING GROUND TARGETS

Paragraphs

SECTION I. General _____ 122–123
    II. Moving vehicles _____ 124–126
    III. Moving personnel _____ 127–128
    IV. Moving targets and ranges and range precau-
        tions_____ 129–130

SECTION I

GENERAL

■ 122. GENERAL.—Personnel armed with the Browning automatic rifle, caliber .30, M1918, will be trained to fire at moving targets, such as tanks, armored vehicles, trucks, and personnel at appropriate ranges. To this end they must be trained in the technique of such fire.

■ 123. FUNDAMENTALS.—*a.* The fundamentals of shooting as presented in chapter 2 apply to firing at moving targets. In applying these fundamentals the firer must adjust his aim and trigger squeeze to the movement of the target.

b. Effective range.—While under exceptional conditions moving targets may be engaged by riflemen armed with the automatic rifle, caliber .30, M1918, at ranges above 600 yards, effective results beyond that range are considered exceptional. Fire at moving targets is, however, usually opened at ranges under 600 yards, and training in the technique of fire is normally limited to firing at such ranges.

c. Leads.—Targets which cross the line of sight at any angle are classified as crossing targets. In firing at such targets the firer must aim ahead of the target so that the paths of the target and the bullet will meet. The distance ahead of the target is called the *lead*. Targets which approach directly towards the firer or recede directly from the firer will for all practical purposes require no lead.

SECTION II

MOVING VEHICLES

■ 124. DETERMINATION AND APPLICATION OF LEADS.—*a.* The lead necessary to hit a moving vehicle is dependent upon the

128

speed of the target, the range to the target, and the direction of movement with respect to the line of sight. Moving at 10 miles an hour, a vehicle moves approximately its own length of 5 yards in 1 second. A rifle bullet moves 400 yards in about one-half second and about 600 yards in about 1 second. Therefore, to hit a vehicle moving at 10 miles an hour at ranges of 400 yards and 600 yards, the leads should be 2½ yards and 5 yards, respectively. At a speed of 20 miles an hour the leads should be 5 yards and 10 yards, respectively.

b. Leads are applied by using the length of the target as it appears to the firer as the unit of measure. This eliminates the necessity for corrections due to the angle at which the target crosses the line of sight because the more acute the angle the smaller the target appears and the less lateral speed it attains. The following lead table is furnished as a guide:

Target length leads

Target speed in miles per hour	For ranges of 400 yards or less	For ranges of 400 to 600 yards
10	½	1
20	1	2

c. As an average rule troops should be instructed to use a lead of 2 target lengths in firing at fast moving targets and of 1 target length against targets which appear to move slowly or to follow an interrupted course.

■ 125. Technique of Fire.—The following technique is employed for firing at moving targets at ranges of 600 yards or less. The battle sight is used. Corrections for range are made by aiming at the center of the target at 600 yards and bringing the aim down to the bottom of the target for ranges of 400 yards or less.

a. *Approaching or receding targets.*—The firer holds his aim on the target in firing each shot.

b. *Crossing targets.*—The firer swings his line of sight through the target and out to the estimated lead. The rifle is kept swinging ahead of the target at the prescribed lead in firing each shot.

c. Fire is executed as rapidly as proper aiming and pressing of the trigger will permit.

■ 126. PLACE IN TRAINING.—The technique of firing at moving vehicles with service ammunition follows training in known-distance marksmanship (ch. 2). When time and ammunition allowances permit, 1,000-inch firing or caliber .22 firing may be added as preliminary instruction.

SECTION III

MOVING PERSONNEL

■ 127. METHOD OF AIMING.—*a.* An elaborate system of calculating leads or of setting sights is neither necessary nor desirable. The following general rule is used with the battle sight. When firing at a man walking across the line of fire the point of aim at the various ranges is taken as follows:

(1) At ranges of 0 to 200 yards, aim directly at the lower part of the body.

(2) At ranges greater than 200 yards, aim at the lower part of the body and lead him the width of his body.

b. When firing at a man advancing or receding from the firer with the battle sight choose a point of aim as indicated in paragraph 125*a.*

■ 128. PLACE IN TRAINING.—As in the case of practice in firing at moving vehicles, instruction in this type of firing should follow instruction in known-distance firing and should immediately precede the training of the squad in the technique of fire.

SECTION IV

MOVING TARGETS AND RANGES AND RANGE
PRECAUTIONS

■ 129. MOVING TARGETS AND RANGES.—*a. Firing at moving vehicles.*—(1) *Target.*—A sled of the type shown in figure 40 has proved to be the most satisfactory kind of target. It has the advantage of a low center of gravity which prevents upsetting on rough ground and in making changes of direction. The sled shown in the figure is 5 by 2½ by 4 feet high and weighs only 45 pounds. Figure 41 shows a similar sled covered with target cloth.

(2) *Towing.*—For towing the target a ½-inch rope has been found satisfactory, the power being furnished by a 1½-ton truck. The pulley shown in figure 41 is simply a channel wheel bolted to a short length of 2-inch board. This board is staked to the ground at a point where a change of direction of the target is desired. The knot shown in the figure should be 10 or 12 feet from the sled, depending on

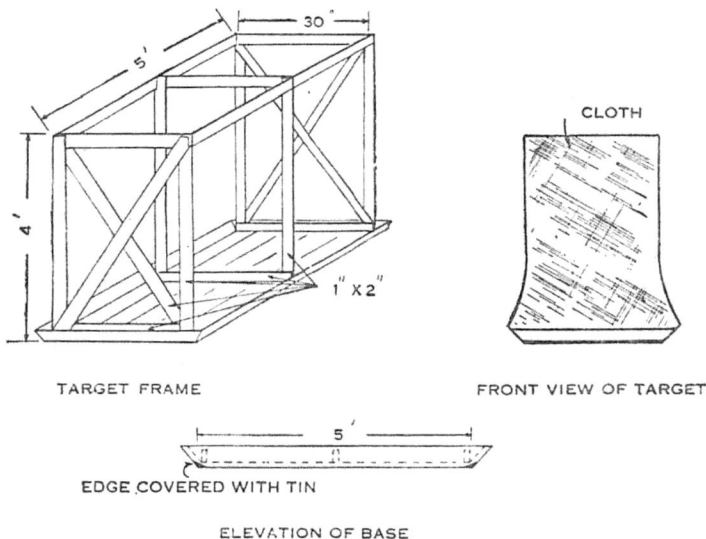

TARGET FRAME   FRONT VIEW OF TARGET

EDGE COVERED WITH TIN

ELEVATION OF BASE

FIGURE 40.—Target frame for moving target.

the speed at which the target is to be run. At faster speeds the knot must be at a greater distance from the sled to prevent the increased momentum of the sled from over-running the pulley.

(3) *Set-up.*—With 500 yards of rope, a set-up as shown in figure 42 can be made. This set-up is only one of the many which it is possible to make with 500 yards of rope. Accidents incident to wrong laying may be prevented by keeping just in rear of the gun a safety officer whose duty is to see

131

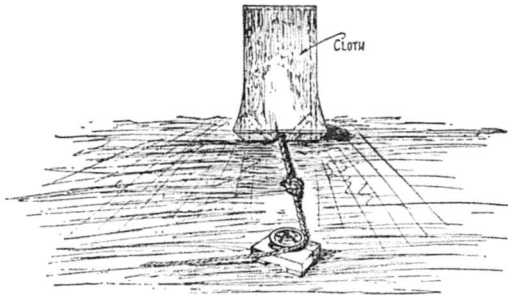

FIGURE 41.—Sled target covered with target cloth; pulley and trip knot for effecting changes of direction.

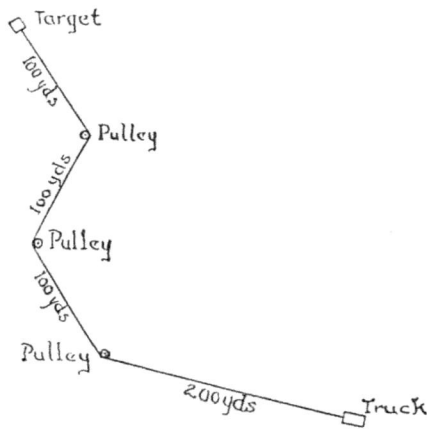

• GUN
FIGURE 42.—Set-up for towing a target.

132

that the barrel is kept pointed in a direction not too near the truck. The essential elements in training a gun squad to fire at moving targets are much practice for the observer in estimating angular speeds and for the gunner in laying on a target in motion, and for everybody, *speed*.

b. Firing at moving personnel.—Any class A range is suitable for this purpose. E targets on sticks carried by men walking in the pits are sufficient.

■ 130. RANGE PRECAUTIONS.—For general range precautions including danger areas, see AR 750–10. In addition to the individual safety precautions prescribed in chapter 2, the following precautions will be observed:

a. Firing at moving targets will not be permitted on any range until the safety angles have been carefully checked and markers have been placed so as to define clearly the right and left limits of fire.

b. Personnel of trucks towing moving targets will operate at such distance from the line of fire as to be protected not only from direct hits but from ricochets.

c. Trucks replacing targets on the course or personnel effecting repairs will be equipped with red flags.

133

CHAPTER 4

MARKSMANSHIP—AIR TARGETS

Paragraphs
SECTION I. Nature of air targets for the automatic rifle____ 131–132
II. Technique of antiaircraft fire_____ 133–137
III. Marksmanship training_____ 138–142
IV. Miniature range practice_____ 143–146
V. Towed-target firing_____ 147–151
VI. Ranges, targets, and equipment_____ 152–157

SECTION I

NATURE OF AIR TARGETS FOR THE AUTOMATIC RIFLE

■ 131. AIR TARGETS SUITABLE FOR AUTOMATIC RIFLE FIRE.—Combat arms take the necessary measures for their own immediate protection against low-flying hostile aircraft. Therefore all troops must be fully trained and imbued with the determination to protect themselves against hostile aerial attacks without reliance upon other arms. All low-flying hostile airplanes are suitable targets for automatic rifle fire.

■ 132. CLASSIFICATION OF AIR TARGETS.—From the point of view of the rifleman air targets may be classified as—

a. *Overhead.*—Those which pass over or nearly over the rifleman.

b. *Nonoverhead.*—Those which do not pass over or nearly over the rifleman.

c. Either of the types listed above may be flying at a constant altitude, or may be decreasing or gaining in altitude.

d. *Direct-diving.*—Targets which dive directly toward a rifleman are called *direct-diving targets.*

e. *Direct-climbing.*—Targets which climb directly away from a rifleman are called *direct-climbing targets.*

SECTION II

TECHNIQUE OF ANTIAIRCRAFT FIRE

■ 133. GENERAL.—Airplanes present very fleeting targets and must be engaged promptly by all available weapons. This

134

section on the technique of fire deals with placing automatic rifle fire on hostile low-flying planes. Details of antiaircraft marksmanship training which deal with firing on various types of targets are contained in sections III, IV, and V.

■ 134. LEADS.—*a. General.*—In order to hit an airplane in flight, the firer must aim ahead of the target so that the paths of the bullet and target will meet. The distance ahead of the airplane is called the *lead*. A lead must be applied in all firing except when the target is at an extremely close range (100 feet), or when it is diving directly at the firer or climbing directly away from him.

b. Application of leads.—The length of the target *as it appears to the firer* is used as the unit of measure for applying leads. The number of times the firer applies this unit of measure is explained in paragraph 136.

■ 135. TARGET DESIGNATION.—*a.* Attacking aviation will often fly in V-shaped formations of three or more airplanes each. Therefore each one of the three airplanes of a typical hostile flight may be designated as a target for an element of the unit or team.

b. Automatic riflemen will assume the firing position as soon as possible after receiving the warning of the approach of hostile airplanes and track the target until it comes within range.

■ 136. INDIVIDUAL TECHNIQUE OF ANTIAIRCRAFT FIRE.—*a.* For all direct-diving or direct-climbing planes, *aim and fire each shot at the target.*

b. For all targets except direct-diving or direct-climbing planes, *aim and fire each shot with a lead of four target lengths.* The target considered in determining the lead of four target lengths is a 30-foot airplane. In using this method for towed-target firing the lead will have to be changed in accordance with the length of the sleeve target.

c. No attempt is made to use the peep sight. Fire by sighting over the top of the rear sight and front sight.

■ 137. DELIVERY OF FIRE.—*a.* Hostile flights are fired upon whenever they come within a range of 600 yards. Fire is maintained while they are within such range unless successive flights appear, in which case fire is not delivered on receding targets.

b. Each shot is aimed and the trigger squeezed. A well-trained automatic rifleman can fire at the rate of about one shot in 2 seconds. Fire is not permitted at faster rates than will permit careful aiming and trigger squeeze.

c. Automatic rifle fire is a serious hazard to low-flying planes and if unhesitatingly delivered will tend to discourage such missions. Hits will frequently cause the airplane to crash and even if the effect of hits cannot be immediately observed may have caused serious damage.

d. The four target length lead prescribed in paragraph 136*b* is suitable for firing on hostile airplanes which have a speed of around 200 miles per hour. This lead should be proportionately increased for hostile airplanes having much greater speed. See paragraph 224*c*.

SECTION III

MARKSMANSHIP TRAINING

■ 138. GENERAL.—*a. Object of instruction.*—*The object of* automatic rifle antiaircraft marksmanship instruction is to train the automatic rifleman to fire effectively at rapidly moving aerial targets.

b. Basis of instruction.—(1) Prior to instruction in antiaircraft marksmanship the automatic rifleman should have completed known-distance marksmanship (ch. 2) and his firing at moving ground targets (ch. 3). To become a good antiaircraft marksman he must be able to apply the fundamentals of marksmanship to firing at rapidly moving targets and to perform the following operations with accuracy and precision:

(*a*) Apply the length of the target as a unit of measure in measuring the required lead.

(*b*) Aline the sights of the rifle at the required lead rapidly by sighting over the top of the rear sight and front sight.

(*c*) Swing the rifle with a smooth, uniform motion so as to maintain the aim on the required lead while squeezing the trigger and during the forward motion of the bolt.

(d) Properly apply the trigger squeeze so as to get the shot off in a minimum of time and without disturbing the aim.

(2) The course of training outlined in this section is intended to train the soldier to obtain the correct performance of the four operations combined into one continuous smooth motion when firing in any direction at rapidly moving aerial targets.

c. *Sequence of training.*—Antiaircraft automatic rifle marksmanship is divided into preparatory exercises, miniature range practice, and towed-target firing.

d. *Personnel to receive training.*—Officers and men as covered in AR 775-10 will receive antiaircraft marksmanship training.

■ 139. PREPARATORY EXERCISES.—a. *General*—(1) *Description.*—The preparatory exercises consist of the following three distinct steps which should be completed on each of the targets described hereafter prior to firing on those targets:

(a) Position exercise.

(b) Aiming and leading exercise.

(c) Trigger-squeeze exercise.

(2) *Method.*—The coach-and-pupil method should be carried on throughout the training. In the preparatory exercise each coach should observe and correct his pupil to see that the following points, as applicable, are observed:

(a) Proper position is taken.

(b) Slack is taken up promptly and firmly.

(c) Rifle is swung with a smooth, uniform motion.

(d) Rifle is swung by pivoting the body at the waist.

(e) Arms, shoulder, rifle, and head move as a unit as the rifle is swung.

(f) Pressure on the trigger is applied promptly, decisively, and continuously.

(g) Eye is kept open and does not blink on the forward motion of the bolt.

(h) Muzzle does not jerk coincident with the release and forward motion of the bolt.

(i) Pupil continues the aim and trigger pressure during the entire length of travel of the target.

b. Organization.—With the targets and target range described hereinafter (see sec. VI), a group of 32 men per target is the most economical training unit. This group is formed in two ranks of 16 men each. For the preparatory exercises this will permit 16 men to perform the exercises on each type of target while the remaining 16 men act as coaches. (See fig. 43.)   When firing the Browning automatic rifle, caliber .30 M1918, the interval between individuals on the firing line should be increased.   This is accomplished by plac-

①Nonoverhead target.

②Overhead target.

FIGURE 43.—Organization for training.

138

ing only one-half the group on the firing line at one time. Each group will complete all preparatory training and instruction firing on its assigned target. Groups then change places. The preparatory training and instruction firing is then undertaken on the new type of target. This procedure is followed until each man of each group has completed his instruction on each of the four types of targets.

■ 140. FIRST STEP—POSITION EXERCISES.—*a*. The kneeling or standing position is generally used in antiaircraft firing.

b. (1) These antiaircraft firing positions differ from those used in firing at ground targets in that—

(*a*) The sling is not used.

(*b*) The arms move freely in any direction with the body.

(*c*) The hands grasp the piece firmly.

(*d*) The butt of the rifle is pulled hard against the shoulder with the right hand and the cheek is pressed against the stock.

(*e*) In the kneeling position the buttock does not rest on the heel, and the left foot is well advanced to the left front. The weight is slightly forward.

(2) The positions are such that the rifle, the body from the waist up, the arms, and the head can move as a unit.

(3) When leading a target the rifle is swung with a smooth, uniform motion. This is accomplished by pivoting the body at the waist. There is no independent movement of the arms, the shoulders, the head, or the rifle.

(4) The instructor explains and demonstrates the position. He points out that if the rifle is pulled or pushed in the desired direction by means of the left hand and arm a jerky motion instead of the smooth swing necessary for correct aiming and trigger squeeze will result.

(5) Position exercises are conducted so that the automatic rifleman will become proficient in rapidly assuming positions for firing at hostile aircraft moving in any direction.

■ 141. SECOND STEP—AIMING AND LEADING EXERCISES.—*a*. *Purpose*.—The purpose of the aiming and leading exercises is to teach the correct method of aiming and to develop skill in swinging the rifle with a smooth, uniform motion.

b. *Method*.—(1) For the instruction of the groups assigned to the nonoverhead targets (see fig. 44 ①) the pupils, in the kneeling firing position, are placed in one line at about a 1½-yard interval, 500 inches from and facing the assigned target. The coaches take position so they can observe the pupils. The commands for the exercise are: 1. AIMING AND LEADING EXERCISE, 2. ONE (TWO) TARGET-LENGTH LEADS, 3. TARGETS. At the command TARGETS, the targets are operated at a speed of from 15 to 20 feet per second. Each pupil alines his sights on the spotter indicating the proper lead, and takes up the slack in his trigger. He then swings his rifle with a smooth, uniform motion by pivoting the body at the waist, and maintains the aim on the proper lead during the travel of the target. The operation is repeated as the target is moved in the opposite direction. The exercise is continued until the target has been moved four times in each direction. The coach and pupil then change places and the exercise is continued until all men have acquired some skill in aiming and leading with one, two, and three target-length leads, both from right to left and left to right.

(2) For the instruction of the group assigned to the overhead target (see fig. 44 ②), the line is formed perpendicular to and facing the line of flight of the target. The procedure is the same except that one target-length lead only is used.

■ 142. THIRD STEP — TRIGGER-SQUEEZE EXERCISES.—*a. General.*—(1) The automatic rifleman is trained to squeeze the trigger exactly as when firing rapid fire at stationary targets except that the rifle is kept in motion during the trigger squeeze, the forward motion of the bolt, the firing of the shot, and momentarily after the firing of the shot.

(2) In firing at a rapidly moving target, the untrained man has a tendency to permit the rifle to come to rest momentarily while applying the final trigger squeeze or during the forward motion of the bolt. This results in the shots going behind the target. Another fault of the untrained man is that of jerking the trigger quickly the instant the aim is on the required lead. This causes a bad shot.

(3) Due to the short period of time during which the usual aerial target will be within effective range, fire is opened as soon as possible and delivered at as rapid a rate as possible

consistent with accuracy. The trigger is therefore squeezed aggressively and decisively.

b. *Object.*—The object of the trigger squeeze exercises is to train the automatic rifleman to apply pressure on the trigger while keeping the rifle in motion, to develop a decisive trigger

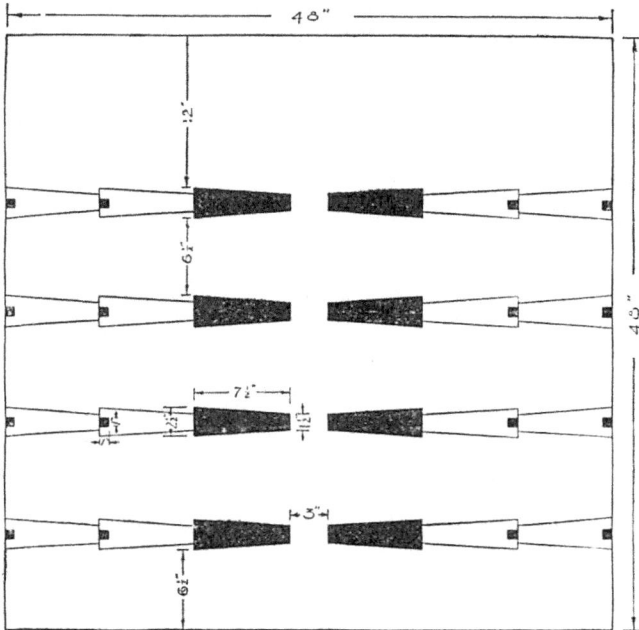

① Nonoverhead.

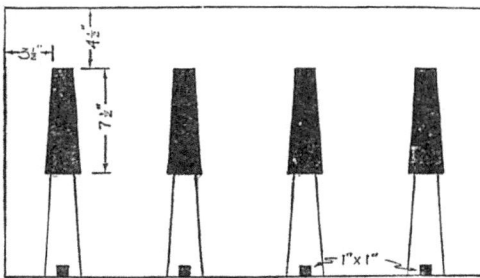

② Overhead.
FIGURE 44.—Aiming and leading targets.

141

squeeze so that fire can be opened in a minimum of time without loss of accuracy, and to train him to follow through with the shot.

c. Method.—(1) The trigger squeeze exercises are conducted in a manner similar to the aiming and leading exercises. The same targets are used but the spotters indicating the various target-length leads will be removed. (See fig. 45.)

(2) The pupils in the kneeling firing position are placed in one line at about a 1½-yard interval, 500 inches from and facing the assigned nonoverhead target. The coaches take position so they can observe the pupils. The commands for the exercise are: 1. SIMULATE LOAD, 2. TRIGGER SQUEEZE EXERCISE, 3. ONE (TWO) TARGET-LENGTH LEADS, 4. TARGETS. At the command TARGETS, the targets are operated at the proper speed. Each pupil takes up the slack in his trigger, estimates the lead announced in the order, applies that lead by swinging the rifle in the manner learned in the aiming and leading exercises, and maintains his aim at the proper lead while applying a constantly increasing pressure on the trigger until the bolt is released. The aim and pressure on the trigger are continued during the entire length of travel of the target regardless of release of the bolt. The importance of following through with the shot must be emphasized. It is by this means that men develop the habit of keeping their rifles in motion during the process of firing. All of the steps explained above are performed in one continuous operation. The exercise consists of having the pupil squeeze the trigger each time the target moves across the front. The exercise for each man consists of four passages of the target in each direction. The coach and pupil then change places and the work is continued until all men have become proficient in squeezing the trigger correctly using various target-length leads.

(3) The procedure for overhead trigger-squeeze exercise is the same except that the line is formed perpendicular to and facing the flight of the target and one target-length lead only is used.

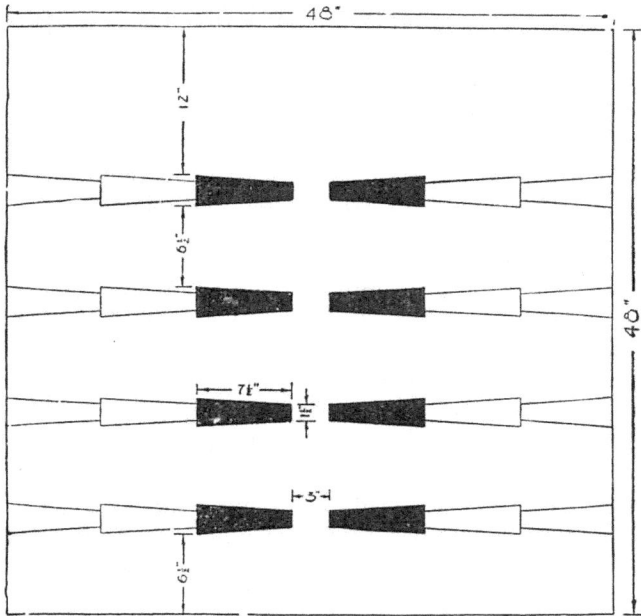

① Nonoverhead.

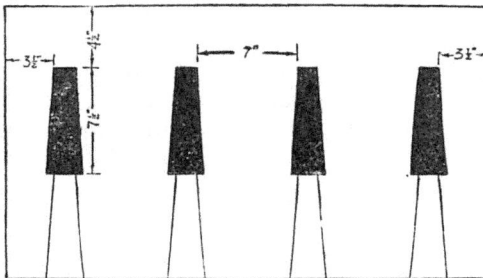

② Overhead.

FIGURE 45.—Instruction targets.

143

SECTION IV

MINIATURE RANGE PRACTICE

■ 143. GENERAL.—*a.* Miniature range practice is divided into two parts—instruction firing and group firing. There is no record firing.

b. All firing is on moving targets on the 500-inch range. A suggested arrangement of the range is given in section VI. Provision is made for simultaneous firing by separate groups on the horizontal, the diving, the climbing, and the overhead targets.

c. The course is fired with the Browning automatic rifle, caliber .30, M1918, if ammunition and danger area permit. If not, the bolt action U. S. rifle, caliber .22, M1922M1 or M1922M2 will be used.

d. All rifles should be zeroed before range practice starts.

■ 144. SAFETY PRECAUTIONS.—*a.* The safety precautions given in paragraph 88 are applicable to this firing and will be observed.

b. If firers are permitted to go forward to inspect their targets, rifles will be left on the firing line. If the Browning automatic rifle, caliber .30, M1918, is used, bolts will be in their forward position and magazines withdrawn. If the caliber .22 rifle is used, bolts will be left open and clips withdrawn.

c. Target operators remain behind the protective wall except when ordered to leave by the officer in charge of the target which they are operating.

d. If the caliber .22 rifle is used, the bolt will not be forced home if difficulty in feeding is experienced. Attempting to force the bolt home may result in igniting a rim-fired cartridge before the cartridge is chambered.

■ 145. INSTRUCTION FIRING.—*a. General.*—(1) The purpose of instruction firing is to teach the soldier to apply the fundamentals taught in the preparatory exercises to actual firing.

(2) During instruction firing each soldier works under the supervision of a coach.

(3) As a group completes the preparatory training on a target, instruction firing is taken up on that target and completed before the group moves to another target.

(4) Instruction firing consists of that indicated in table XVI, which follows *b*(13) below.

b. Procedure.—(1) As the instruction firing on each type of target follows immediately after the preparatory exercises on that target, the organization of the training unit for firing should be the same as that given in paragraph 139*b*.

(2) The front rank of each group is formed on the firing line *in the kneeling firing position*. The rear rank men act as coaches.

(3) One-half of the front rank of the group fires while the remaining front rank men simulate firing.

(4) A silhouette is assigned to each individual firer. For example, the four silhouettes on the right of the target are assigned the first four men on the right of the line; the four silhouettes on the left of the target are assigned the next four men. Silhouettes for the men simulating firing are assigned in the same manner, that is, the right four are assigned silhouettes on the right of the target and the left four are assigned silhouettes on the left of the target.

(5) The officer in charge of the target commands: 1. LOAD, 2. ONE (TWO) TARGET-LENGTH LEADS, 3. TARGETS. At the command TARGETS the targets are operated at the proper speed. Men assigned silhouettes on the right half of the nonoverhead target aim and fire in accordance with the method learned in the trigger-squeeze exercise (par. 142*c*(2)). They fire one shot each time the target crosses from their right to left. Men assigned silhouettes on the left half of the nonoverhead target fire one shot each time the target crosses from left to right.

(6) Men assigned silhouettes on the overhead target fire one round each time the target is run in the approaching direction in the same manner as explained above.

(7) Four rounds constitute a score. After each string of four rounds, targets are scored and shot holes penciled.

(8) One point is awarded for each hit in the silhouette when using one target-length lead, or in the proper scoring space when using more than one target-length lead.

(9) Half-groups alternate firing and simulating firing.

(10) When front rank men have fired one score as the target has moved in each direction, they change places with the men in the rear rank and act as coaches.

(11) This procedure is followed until all men of the group have performed the required firing.

(12) Upon completion of the firing prescribed in table XVI below for any one type of target, the group moves to another type target and continues until all have completed the instruction firing.

(13) Modifications of the above method of firing to meet local conditions are authorized.

TABLE XVI.—*Instruction firing*

(Range 500 inches)

Target	1 lead, 8 rounds	2 leads, 8 rounds	3 leads, 8 rounds	Total
Horizontal	4 rounds right to left. 4 rounds left to right.	4 rounds right to left. 4 rounds left to right.	4 rounds right to left. 4 rounds left to right.	24
Climbing	4 rounds right to left. 4 rounds left to right.	4 rounds right to left. 4 rounds left to right.	4 rounds right to left. 4 rounds left to right.	24
Diving	4 rounds right to left. 4 rounds left to right.	4 rounds right to left. 4 rounds left to right.	4 rounds right to left. 4 rounds left to right.	24
Overhead	4 rounds approaching 4 rounds receding	4 rounds approaching 4 rounds receding		16

Speed of all targets, 15 to 20 feet per second. Total rounds, 88.

■ 146. GROUP FIRING.—*a. General.*—(1) Group firing is the final phase of antiaircraft marksmanship training on the miniature range.

(2) It provides for competitions and illustrates the effectiveness of the combined fire of a number of automatic riflemen.

(3) Group firing should not be undertaken until the preparatory training and instruction firing have been completed.

b. Procedure.—(1) Two silhouettes, one to be fired upon as the target moves from left to right and one to be fired upon as the target moves in the opposite direction, are assigned to each squad or similar group.

(2) Each man of the front rank followed by each man in the rear rank fires four rounds at each silhouette as the target moves in the appropriate direction.

(3) Targets are scored after completion of the firing of the entire unit or group.

c. Scoring.—A value of 1 is given each hit on the silhouette.

d. Score card.—A sample score card is shown in paragraph 153*e*.

SECTION V

TOWED-TARGET FIRING

■ 147. GENERAL.—*a.* Towed-target firing is the final phase of automatic rifle antiaircraft marksmanship. It is conducted on the towed-target range described in section VI.

b. It consists of firing with caliber .30 ball or tracer ammunition at a sleeve target at various ranges and on varied courses.

c. Towed-target courses prescribed herein are guides which may be modified. Safety measures and ammunition requirements restrict the length of the course. Safety measures also prevent the adoption of courses to include direct-diving or direct-climbing targets.

d. Towed-target firing will follow miniature range instruction firing. If, due to lack of facilities, a unit is unable to conduct miniature range firing, it may be permitted to conduct towed-target firing, provided antiaircraft marksmanship preparatory training has been completed.

■ 148. COURSES TO BE FIRED.—Units authorized to fire will fire one or more of the courses enumerated in table XVII below.

147

TABLE XVII.—*Courses to be fired*

Course No.	Type of flight	Altitude of target	Horizontal range of course (yards)[1]	Speed	Remarks
1	Nonoverhead— horizontal (parallel to firing line from left to right).	Minimum consistent with safety.	Minimum, 100; maximum depends on width of danger area of range.	Maximum possible.	See figure 57.
2	Nonoverhead— horizontal (parallel to firing line from right to left).	----do----------	----do----------	----do------	Do.
3	Overhead (perpendicular to firing line).	----do----------	Minimum and maximum in accord with safety precautions.	----do------	See figure 58.
4	Combined courses 1, 2, and 3.	----do----------	Same as for courses 1, 2, and 3.	----do------	See figure 59.

[1] The horizontal distance from the firing point directly under the target. The maximum slant range for all courses should not exceed 600 yards.

■ 149. SAFETY PRECAUTIONS.—*a.* Towed-target firing will be conducted with due regard for the safety of the pilot of the towing airplane, the personnel engaged in firing, and all spectators.

b. All firing is controlled by suitable signals or commands. COMMENCE FIRING and CEASE FIRING are given in such a manner as to be clearly and promptly understood by everyone engaged in firing.

c. The signals and commands COMMENCE FIRING and CEASE FIRING will be given at such time as to prevent any bullets from falling outside the danger area.

d. For all overhead flights, the signal or command COMMENCE FIRING will not be given until the towing plane has reached a point 50 yards or less, measured horizontally on the ground, from the firing point and there is no danger of bullets striking the plane. The signal or command CEASE FIRING will be given when the sleeve target is at least 100 yards, measured horizontally on the ground, in advance of the firing line so there is no danger of bullets dropping outside the firing area.

e. Whenever a towing cable breaks and the towing airplane is on a course which passes near the firing point, all personnel in that vicinity will be warned to lie flat on the ground until danger from the loose cable and the release is past.

f. No rifle will be pointed at or near the towing airplane. All tracking will be on the towed target. Muzzles will be depressed during loading.

g. At least two safety officers will be designated to assist the officer in charge of firing in carrying out safety precautions.

h. To provide for the safety of the towing airplane, firing will be permitted only when the smaller angle in space between the gun-target line and the towline (or towline extended) is greater than 45°.

i. An Air Corps officer should be at the firing point during an organization's initial practice for the season, for the purpose of giving supplemental instruction and checking the safety measures taken.

j. Additional safety precautions are covered in AR 750–10.

■ 150. PROCEDURE OF FIRING.—*a.* The men to fire take the antiaircraft kneeling firing position on the firing line with at least 1½ yards between men.

b. The officer in charge of firing takes position in rear of the center of the firing line.

c. Safety officers take position at either flank of the firing line.

d. As the *towing airplane* approaches the left (right) side of the danger area, the officer in charge of firing gives the command 1. (SO MANY) ROUNDS, 2. LOAD, 3. SLEEVE TARGET APPROACHING FROM THE LEFT (RIGHT). Each automatic rifleman loads his piece and sets it at "safe."

e. As the *towed target* approaches the danger area, the officer in charge of firing commands 4. FOUR TARGET-LENGTH

LEADS. (See par. 213c.) At this preparatory command each automatic rifleman sets his piece at *F*, aims by swinging through the sleeve to the announced lead, pivoting at the waist, and maintains his estimated lead.

f. In firing at crossing targets, the safety officer stationed at the end of the firing line opposite to the target's approach signals or commands COMMENCE FIRING when the sleeve-target has completely crossed the line marking the firing area. The officer in charge of firing and such assistants as he desires repeat the command or signal to insure that all firers hear it. Each automatic rifleman squeezes the trigger until the bolt is released and the first shot is fired. He then continues to re-aim rapidly and fire until the command or signal CEASE FIRING is given. The safety officer at the end of the firing point opposite to the target's departure observes the flight of the sleeve-target during the firing. When he observes that the sleeve is about to leave the danger area he signals or commands CEASE FIRING. The officer in charge of firing and his assistants repeat the command or signal to insure that all firers hear it.

g. In firing at overhead targets the same procedure is followed except that the officer in charge of firing, from his position behind the center of the firing line, determines when firing commences and ceases. He gives the command or signal COMMENCE FIRING when the towing plane is 50 yards or less in advance of the firing line and gives CEASE FIRING before the sleeve is 100 yards in advance of the firing line. (See par. 149.)

■ 151. SCORING.—*a*. The number of hits is found by dividing the number of holes in the target by two. An odd hole is counted as a hit.

b. The hit percentage is obtained by dividing the number of hits as obtained in *a* above by the total number of rounds fired at the target.

SECTION VI

RANGES, TARGETS, AND EQUIPMENT

■ 152. RANGE OFFICER.—A range officer is appointed well in advance of range practice. His chief duties are stated in paragraph 80c.

■ 153. MINIATURE RANGE.—*a*. The miniature range consists of:

(1) One horizontal target. (See fig. 46.)

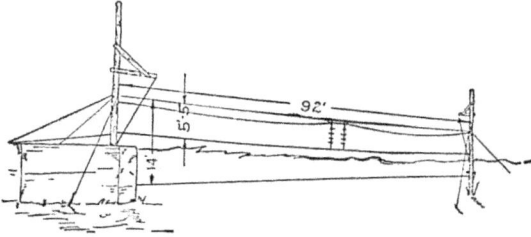

FIGURE 46.—Horizontal target.

(2) One double-climbing and diving target. (See fig. 47.)

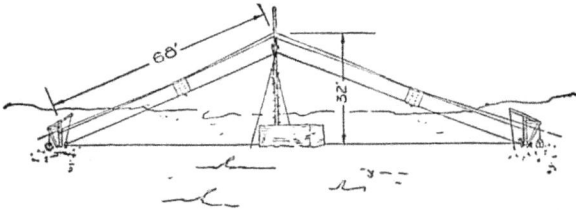

FIGURE 47.—Double-climbing and diving target.

(3) One overhead target. (See fig. 48.)

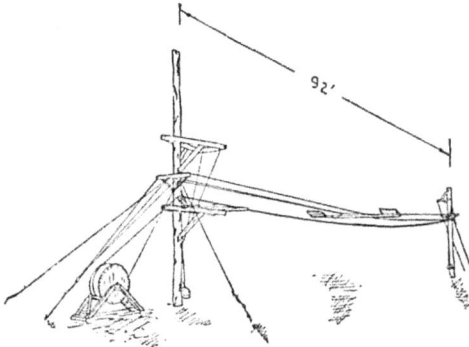

FIGURE 48.—Overhead target.

b. A suggested arrangement of the targets is shown in figure 49.

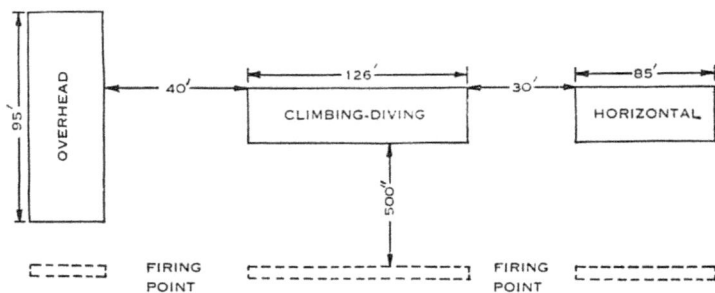

FIGURE 49.—Arrangement of targets.

c. For details of range apparatus, see figures 51 to 56, inclusive, which are included at the end of this section.

d. Danger area required.—(1) The danger area required is dependent upon the type of ammunition. (See AR 750–10 for size and shape.)

(2) The miniature range may be laid out in the same manner as described in paragraph 154c. Care must be taken to insure that the firing line and targets are placed so that no fire will fall outside of the danger area.

e. Equipment required.—If the organization for training is as suggested in paragraph 139b, the following equipment is necessary:

1 Browning automatic rifle, caliber .30, M1918, for each two men firing, or 1 caliber .22 rifle for each two men firing.

4 aiming and leading targets (see fig. 44). (Each of these targets consists of a piece of beaverboard on which the silhouettes are pasted.)

6 instruction firing targets per range (see fig. 45). (These targets are the same as the aiming and leading targets except that the spotters are eliminated.)

1 score card per man as follows:

INDIVIDUAL SCORE CARD

ANTIAIRCRAFT RIFLE MARKSMANSHIP

Date................, 19....

Name....................

Target	1 TL lead			2 TL lead			3 TL lead		
	Rounds fired	Hits	Per-cent	Rounds fired	Hits	Per-cent	Rounds fired	Hits	Per-cent
Horizontal.........									
Climbing..........									
Diving...........									
Overhead.........									
Total........									

▉ 154. TOWED-TARGET RANGE.—*a*. In selecting the location of a towed-target range, the danger area is the chief consideration. (See AR 750–10.)

b. The firing point should accommodate at least 50 men in line with a 1½-yard interval between men. A level strip of ground, preferably on a hill, 75 yards long and 2 yards wide is suitable. A firing point similar to the firing point of a known-distance rifle range may be built (see par. 121c(11)).

c. (1) After the towed-target range has been selected, the firing point, limits of fire, and danger area should be plotted on a map or sketch of the area.

(2) From this map or sketch the range is then laid out on the ground. First, each end of the firing point is marked by a large stake. The right and left limits of fire are then each marked by a post. Each post is placed at the maximum distance at which it will be plainly visible from the firing point. When these distances have been determined, the posts are located in azimuth by the following method: To locate the post marking the left limit of fire, an aiming circle or other angle-measuring instrument is set up at the right end

stake of the firing point. It is then oriented and laid on an azimuth which, by reference to the map or sketch, is known to be the farthest to the left that the rifle at the right end of the firing point can safely be fired. The post marking the right limit of fire is similarly located with the instrument set up at the left end stake of the firing point. (See fig. 50.)

(3) Direction guides for the towing airplane to follow should, within the limits of fire, be distinctly marked on the ground for each course. White targets or strips of cloth placed flat on the ground about 30 feet apart are suitable.

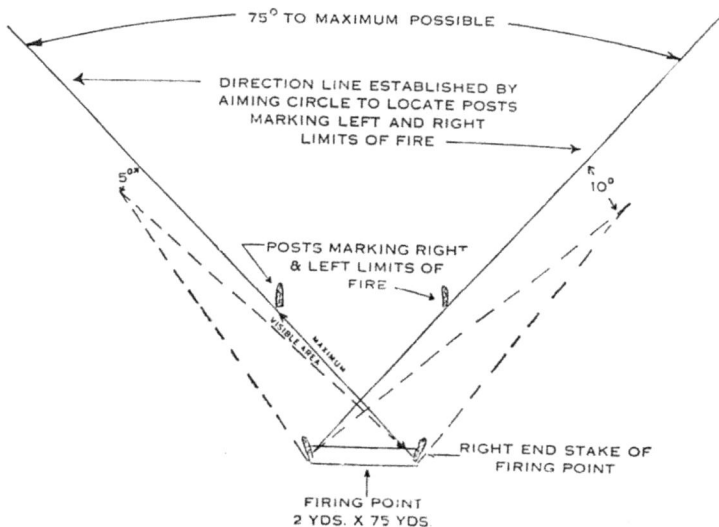

FIGURE 50.—Towed-target range showing firing point and limit of fire. Dotted lines show danger area.

■ 155. TOWED TARGETS.—*a. Type and source.*—The targets used in towed-target firing are sleeve-targets furnished by the Air Corps unit assigned the towing mission. They are returned to the Air Corps unit after they have been scored.

b. Towline.—The towing line will be not less than 600 yards long.

■ 156. INSTRUCTION TO PILOTS FOR TOWING MISSIONS.—*a.* Towed-target firing requires the closest cooperation between

154

the pilot of the towing airplane and the officer in charge of firing. Decisions affecting the safety of the plane rest with Air Corps personnel.

b. The air mission for towed-target firing should be specifically stated. The commanding officer requesting airplanes for towed-target firing should furnish, in writing, to the Air Corps unit commander concerned the following information:

(1) Place of firing.

(2) Date and hour of firing.

(3) Number of missions to be flown; altitude, course, speed, and number of runs for each.

(4) Ground signals to be used.

(5) Map of the area with the firing line, angle of fire, danger area, the course of each mission, and the location of the grounds for dropping targets and messages all plotted thereon. An alternate dropping ground should be designated when practicable and either or both dropping grounds are subject to approval by the pilot.

(6) Length of towline, within limits established by the Air Corps and subject to approval by the pilot.

(7) Number of sleeve-targets required.

c. Whenever practicable, the officer in charge of the firing should discuss with the pilot the detailed arrangements mentioned in *b* above. This discussion should take place on the towed-target range where the various range features can be pointed out to the pilot. The courses over which the airplane is to be flown should be distinguished on the ground (within the angle of fire). Machine-gun targets placed flat on the ground, about 30 feet apart, or strips of target cloth are practicable for this purpose on some courses. On others a terrain feature such as a beach line may be used.

■ 157. SIGNALS.—*a.* Direct radio communication is the most effective means by which the officer in charge of towed-target firing and the pilot of the towing plane maintain contact with each other. Even though radio is being used, panels should be available in case radio communication fails.

b. For signaling from the ground to the pilot any method agreed upon may be used.   The panel signals generally used are as follows:

Stand by_____ 0 0 2
Ready to fire_____ 0 0 0
Repeat run No. 1_____ 0 9 1
Repeat run No. 2_____ 0 9 2
Repeat run No. 3_____ 0 9 3
Repeat course_____ 0 9 4
Mission complete_____ Pick up panels

c. The pilot may also communicate with the officer in charge of firing by dropped messages or by rocking his wings.

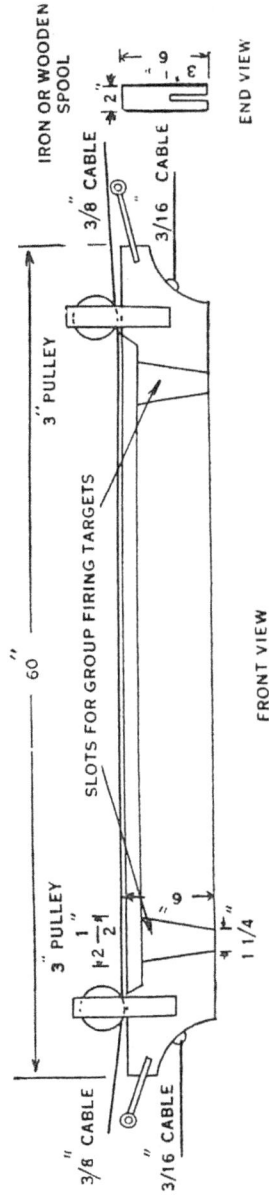

FIGURE 51.—Nonoverhead target carrier.

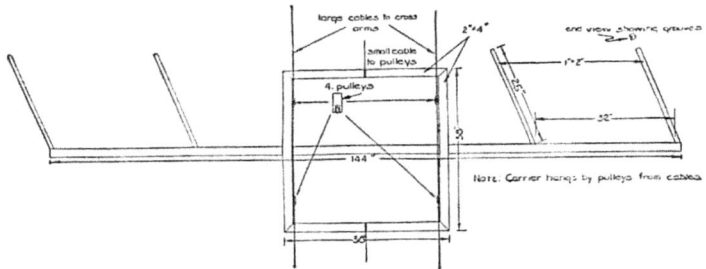

FIGURE 52.—Overhead target carrier.

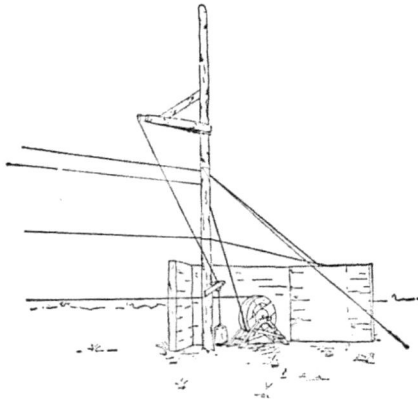

FIGURE 53.—Rear view of nonoverhead range butts, showing drum, guide wires, and bumper.

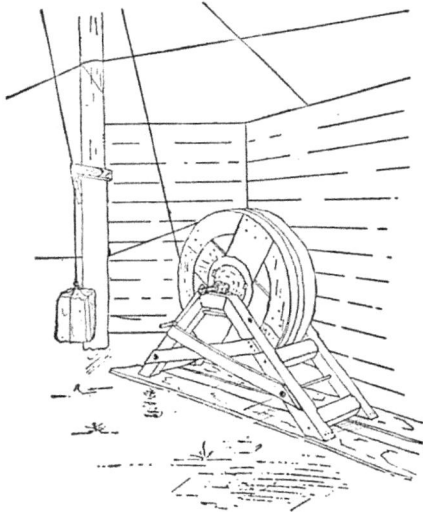

FIGURE 54.—Moving target drum. One complete turn moves target 15 feet.

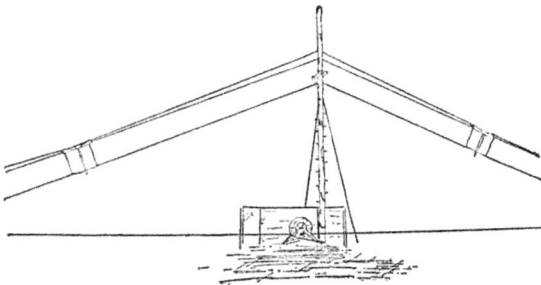

FIGURE 55.—Rear view of climbing and diving target.

159

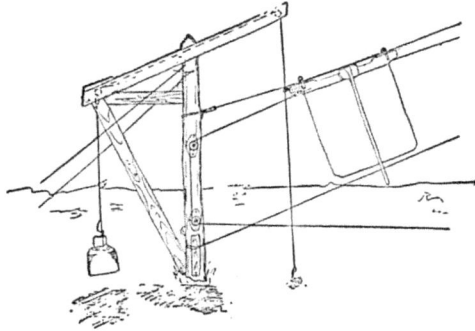

FIGURE 56.—Rear view of climbing and diving target and method of securing target to frame.

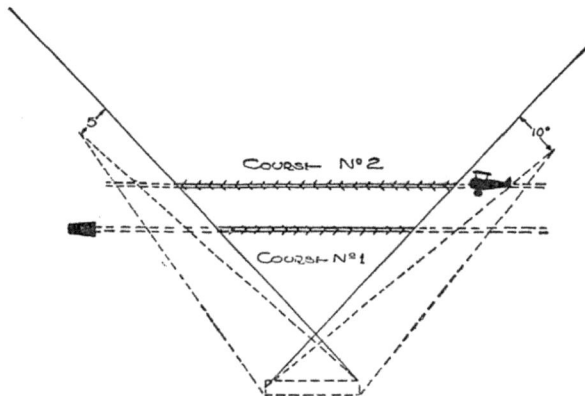

FIGURE 57.—Courses Nos. 1 and 2. Firing takes place when target is on shaded portion of course.

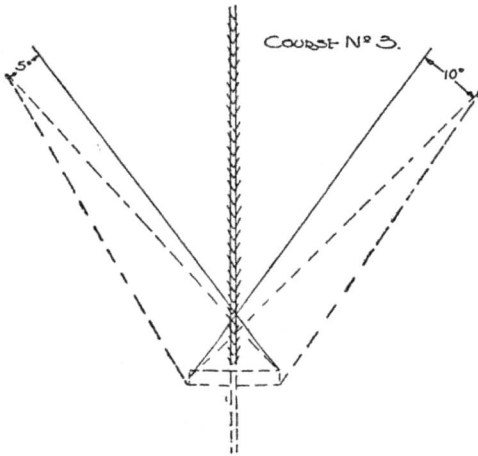

FIGURE 58.—Course No. 3. Firing takes place when target is on shaded portion. Fire is opened when towing airplane is 50 yards or less from firing point.

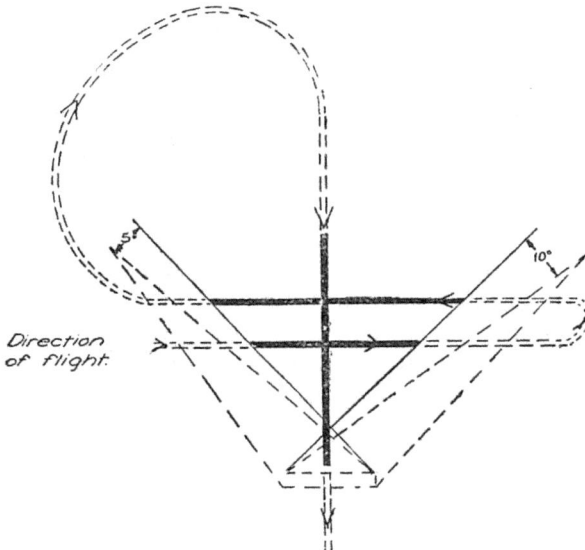

FIGURE 59.—Course No. 4. Heavy lines indicate when towed target is fired upon.

161

CHAPTER 5

TECHNIQUE OF FIRE

                                                              Paragraphs
Section I. Introduction _____ 158–160
        II. Range estimation _____ 161–165
       III. Target designation_____ 166–173
        IV. Rifle fire and its effect_____ 174–180
         V. Application of fire_____ 181–188
        VI. Landscape-target firing_____ 189–196
       VII. Firing at field targets_____ 197–202

Section I

INTRODUCTION

■ 158. General.—*a.* Instruction in the technique of fire is given to automatic riflemen after they have completed or progressed sufficiently in other allied subjects, such as known-distance marksmanship, extended order, drill and combat signals, and certain elements of scouting and patrolling. This chapter deals with instruction in the technique of fire. While the application of this training to combat should be kept in mind, it does not include the solution of tactical exercises.

b. Collective fire is the combined fire of a group of individuals. It may include the fire of several different weapons.

c. A *fire unit* is one whose fire in battle is under the immediate and effective control of its leader. The rifle fire unit is usually the squad or a smaller group.

■ 159. Importance of Rifle Fire.—Effective rifle fire is a characteristic of successful Infantry and is an element which may determine the issue of battle. Collective fire is most effective when it is the product of teamwork.

■ 160. Scope.—Instruction is progressive and is divided into six consecutive steps. These are:

a. Range estimation.

b. Target designation.

c. Rifle fire and its effect.

d. Application of fire.

e. Landscape-target firing.

f. Firing at field targets.

162

SECTION II

RANGE ESTIMATION

■ 161. IMPORTANCE.—*a.* The battle sight for the Browning automatic rifle, caliber .30, M1918, corresponds to a sight setting of approximately 550 yards. With the use of the battle sight a shot will strike the target with the following approximate relation to the point of aim with respect to elevation:

Range (yards)	Strike with use of battle sight
600	13 inches below point of aim.
550	Strikes point of aim.
500	13 inches above point of aim.
400	24 inches above point of aim.
300	23 inches above point of aim.
200	14 inches above point of aim.
100	5 inches above point of aim.

b. It is therefore important for the leader or individual soldier to be able to estimate the range to the target in any circumstances and to decide whether the battle sight or a more exact setting will be used.

■ 162. METHODS.—The following methods of estimating ranges are considered in instruction in the technique of rifle fire:

a. Use of tracer bullets.

b. Observation of fire.

c. Estimation by eye.

■ 163. USE OF TRACER BULLETS.—The leader, or individual, fires a tracer bullet with his sight set at the estimated range. He then corrects the sight-setting according to the strike of the bullet and continues the process until a tracer appears to strike the target. The estimator then announces the correct range, making allowance for the zero of his own rifle.

■ 164. OBSERVATION OF FIRE.—This method can be used with ordinary ball cartridges when the ground is dry and the strike of the bullet can be seen. The same procedure is followed as in determining the range by tracer bullets.

■ 165. ESTIMATION BY EYE.—*a. Necessity for training.*—The usual method of estimating ranges in combat is estimation by eye. Untrained men make an average error of 15 percent of the range when estimating by eye. Hence, a definite system of range estimation, coupled with frequent practice, on varied terrain, is essential to success with this method.

b. Unit-of-measure method.—(1) Ranges less than 500 yards are measured by applying a mental unit of measure 100 yards long. Thorough familiarity with the 100-yard unit and with its appearance on varied terrain and at different distances is necessary if the soldier is to apply it accurately.

(2) Ranges greater than 500 yards are estimated by selecting a point halfway to the target, applying the unit of measure to this halfway point, and doubling the result.

(3) The average of a number of estimates by different men will generally be more accurate than a single estimate. This variation of the suggested method is used, when time permits, by taking the average of the estimates of members of the squad or group or of specially qualified men.

c. Appearance of objects.—If much of the ground between the observer and the target is hidden from view, the application of the unit of measure is impracticable. In such cases the range is estimated by the appearance of objects. Whenever the appearance of objects is used as a basis for range estimation, the observer must make allowances for the following effects:

(1) Objects seem nearer—

(*a*) When the object is in a bright light.

(*b*) When the color of the object contrasts sharply with the color of the background.

(*c*) When looking over water, snow, or a uniform surface like a wheat field.

(*d*) When looking downward from a height.

(*e*) In the clear atmosphere of high altitudes.

(*f*) When looking over a depression most of which is hidden.

(2) Objects seem more distant—

(a) When looking over a depression, most of which is visible.

(b) When there is a poor light or fog.

(c) When only a small part of the object can be seen.

(d) When looking from low ground upward toward higher ground.

d. *Exercises.*—(1) *Exercise No. 1.*—(a) *Purpose.*—To familiarize the soldier with the 100-yard unit of measure.

(b) *Method.*—Units of measure, 100 yards each, are staked out on varied ground, using markers that will be visible up to 500 yards. The men are required to become thoroughly familiar with the appearance of each unit of measure from the prone, kneeling, and standing positions at various ranges.

(2) *Exercise No. 2.*—(a) *Purpose.*—To illustrate the application of the unit of measure.

(b) *Method.*

1. Ranges up to 900 yards are measured accurately and marked at every 100 yards by large markers or target frames, each bearing a number to indicate its range. Men undergoing instruction are then placed about 25 yards to one side of the prolonged line of markers and directed to place a hat or other object before their eyes so as to exclude from view all of the markers. They are then directed to apply the unit of measure five times along a straight line parallel to the line of markers. When they have selected the final point the eye cover is removed and the estimations of the successive 100-yard points and the final point are checked against the markers. Accuracy is gained by repeating the exercise.

2. Ranges greater than 500 yards are then considered. With the markers concealed from view men estimate the ranges to points which are obviously over 500 yards distant and a little to one side of the line of markers. As soon as they have announced each range they remove their eye covers and check the range to the target and to the halfway point by means of the

markers. Prone, sitting or kneeling, and standing positions are used during this exercise.

(3) *Exercise No. 3.* — (*a*) *Purpose.* — To give practice in range estimation.

(*b*) *Method.*—From a suitable point ranges are previously measured to objects within 1,000 yards. The men are required to estimate the ranges to the various objects as they are pointed out by the instructor, writing down their estimates on paper pads or slips. At least one-half of the estimates are made from the prone or sitting positions. Thirty seconds are allowed for each estimate. When all ranges have been estimated the papers are collected and the true ranges announced to the class. To create interest individual estimates and squad averages may be posted on bulletin boards.

SECTION III

TARGET DESIGNATION

■ 166. IMPORTANCE.—Target designation is a vital element in the technique of fire unless the target is self-evident. Battlefield targets are generally so indistinct that leaders and troops must be able to designate their locations and extent. Small units and individuals must also be trained to place heavy fire on indistinct or probable targets in appropriate circumstances.

■ 167. INSTRUCTION.—Prior to instruction in target designation automatic riflemen should understand the topographical terms normally employed in designating targets, for example, crest, military crest, hill, cut, ridge, bluff, fill, ravine, crossroads, road junction, road center, road fork, skyline.

■ 168. METHODS.—The following methods are used to designate targets:

    a. Tracer bullets.
    b. Pointing.
    c. Oral description.

The method used should be the one best suited to the conditions existing at the time of the appearance of the target.

■ 169. TRACER BULLETS.—*a.* The use of tracer bullets is a quick and sure method of designating an obscure battlefield target. Their use, however, has limitations, for they may disclose the position of the firer to the enemy; further, the effect of a sudden burst of fire is lessened by preceding it with tracers.

b. To designate a point target by this method, the leader, or individual, announces, "Range 500; watch my tracer," and fires a tracer at the target. If the target has width, the flanks are indicated by tracer bullets and announced, "Left flank; right flank." Any range correction should be announced.

■ 170. POINTING.—Targets may be pointed out either with the arm or the rifle. Pointing may be supplemented by oral description. To use the rifle for this purpose, it is canted to the right and aimed at the target. The head is then straightened up without moving the rifle. A soldier standing behind looks through the sights and locates the target. If time permits a muzzle rest can be improvised for a rifle aimed at the targets. In pointing, the range is always announced. Usually some supplementary description will be necessary.

■ 171. ORAL DESCRIPTION.—*a. Use.*—Oral description is often used to designate targets. However, battlefield conditions will rarely permit the leader to designate a target directly to all members of his unit by this method. For this reason either pointing or tracers are frequently used in combination with oral description.

b. Elements of oral target designation.—The elements of oral target designation are:

(1) Range.

(2) Direction.

(3) Description of target.

These elements are always given in the above sequence with a slight pause between each element. An exception to this rule occurs when the target is expected to be visible for a short time only. In this case the target is pointed out as quickly as possible; for example, such an oral target designation might be "Those men." No range is announced and men open fire with the sight setting then on their rifles. (Fig. 60, target at K.) If time permits the range is announced and men immediately set their sights before looking for the target.

167

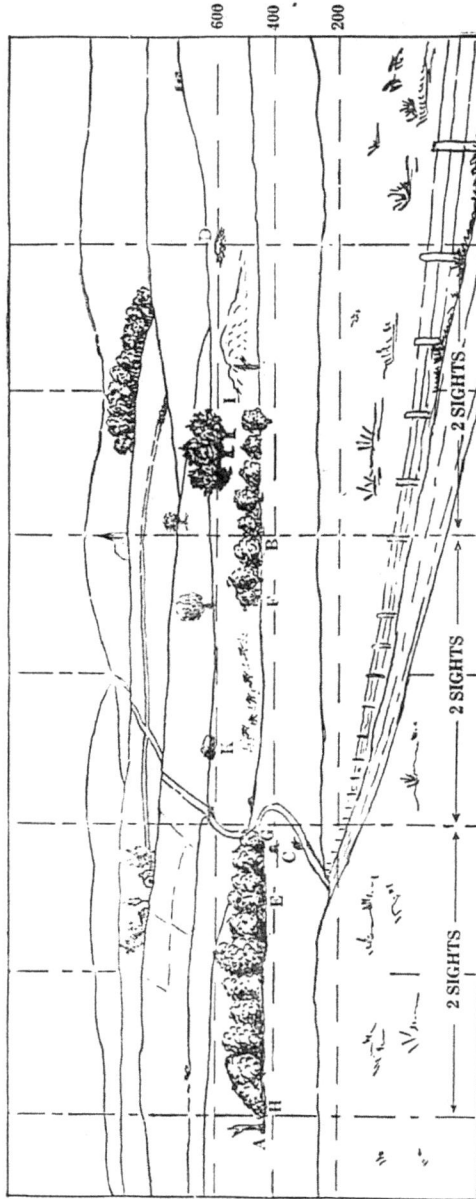

FIGURE 60.—Field targets.

c. Direction.—The terms front, left front, right front, left flank, and right flank, may be used to indicate the general direction of the target. When necessary, the direction is fixed more accurately by the methods hereafter described.

d. Simple description.—When the target is plainly visible, or at an easily recognized point, a simple description is used; for example (fig. 60, target at A) :

Range: 425.

Left front.

Sniper at base of dead tree.

e. Reference point.—When the target is indistinct or invisible, and is not located at some prominent point, the direction of the target is indicated by the use of a reference point. This is an object, preferably a prominent one, by reference to which the location of other points may be determined. In selecting a reference point, care must be taken that another similar object is not mistaken for the one intended. A reference point on a line with the target and beyond it will give greater accuracy than one between the observer and the target. For brevity a reference point is called "Reference."

(1) *When reference point is on line with target.*—The description takes the following form (fig. 60, target at B) :

Range: 450.

Reference: church spire.

Target: machine gun in edge of woods.

It will be noted that the range announced is that to the target and not to the reference point. When the word *reference* is used the word *target* is also used to differentiate between the two objects. Another example follows (fig. 60, target at C) :

Range: 300.

Left front.

Reference: black stump.

Target: sniper on far side of road.

(2) *(a) When reference point is not on line with target.*—In this case it is necessary to indicate the distance to the right or left of the reference point at which the target is located. This distance is measured in units called sights (see par. 173*b*). Suppose that the rifle is pointed so that the left edge of the raised sight leaf is on line with the reference

point and it is found that the right edge of the sight leaf is in line with the target, the target is then one sight width to the right of the reference point and it is announced as "Right, one sight." If the sight can be applied one and one-half times in the above manner, the target is "Right, one and one-half sights." The following examples illustrate this method (see fig. 60):

(Target at D)—
 Range: 600.
 Reference: church spire. Right, two sights.
 Target: group of enemy in shell hole near crest.
(Target at E)—
 Range: 425.
 Left front.
 Reference: dead tree. Right one and one-half sights.
 Target: sniper in edge of woods.
(Target at F)—
 Range: 450.
 Reference: church spire, left one-half sight.
 Target: machine gun in corner of woods.

(b) The width or extent of targets is also measured in sights (fig. 60, target G to H):
 Range: 425.
 Reference: church spire, left two sights.
 Target: enemy groups in edge of woods extending left two sights.

(3) Successive reference points may be used instead of sight measurements from one reference point (fig. 60, target at I). The following example illustrates this method:
 Range: 500.
 Reference: church spire; to the right and at a shorter range, group of three trees; to the right and at the same range.
 Target: machine gun at left end of mound of earth.

(4) *Combination of successive reference points and sights.*—Example (fig. 60, target at K):
 Range: 600.
 Reference: church spire; to the left and at a shorter range, lone tree; left one sight and at the same range.
 Target: machine gun in clump of brush.

f. Variations.—If one end of a linear target is considerably nearer than the other, the average range is announced, since dispersion will cover the target. In oral description the simplest, briefest, and clearest description that fits the conditions is the most effective. Informal or conversational description may be used to supplement the more formal description when the target is not recognized from the latter alone.

■ 172. METHOD TO BE USED.—Troops should be trained in *all* the methods of target designation, that is, *use of tracer bullets, pointing,* and *oral description.* When methods are equally effective, the simplest will be used.

■ 173. EXERCISES.—*a. Exercise No. 1.*—(1) *Purpose.*—To afford practice in target designation by means of tracer bullets.

(2) *Method.*—(a) On a known-distance or field firing range a concealed target representing a machine gun is placed near a pit or other bulletproof shelter. About 500 yards in front of the target a firing position suitable for a squad is selected. The location of the target should be visible from the firing position, but the target itself is carefully concealed.

(b) The unit or team is deployed along the firing position and all except the leader are then faced to the rear.

(c) The leader takes the prone position and is told that the waving of a red flag to his front will represent the firing and smoke from the machine gun.

(d) A man stationed in the pit waves a flag in front of the target for about 30 seconds and retires to the protection of the pit.

(e) The unit or team is faced to the front and men take the prone position. Rifles are loaded, the leader using tracer ammunition and the remainder of the unit or team ball cartridges.

(f) The leader points out the target by firing tracers and announces the range, which is passed orally from man to man.

(g) As soon as each man understands the location of the target he opens fire with the proper sight setting.

(h) Shortly after all the men have taken up the firing the instructor terminates the exercise.

(i) The leader observes the firing. The second in command assists the leader.

(j) After firing ceases, sight settings are checked by the leader and the target is examined or the hits are signaled to the unit or team.

b. Exercise No. 2.—(1) *Purpose.*—To teach the use of sights and fingers for lateral measurement.

(2) *Method.*—(a) A number of short vertical lines one foot apart are plainly marked on a wall or other vertical surface. At a distance of 20 feet from the wall a testing line is drawn or marked out by stakes. The instructor explains that the vertical lines are one sight (50 mils) apart when measured from the testing line, so that the correct distance from the rifle sight leaf to the eye can be determined by pointing the rifle at the vertical lines and moving the eye along the stock until the raised sight leaf covers the space between one of the vertical lines and the next line to the right or left. The instructor demonstrates with a rifle while explaining.

(b) The men take positions on the testing line and each determines the proper distance of his eye from the sight as explained by the instructor. The position of the eye with reference to the stock is carefully noted or marked on the stock. (This will usually be about 14 inches from the eyes.)

(c) The instructor then explains and demonstrates the use of fingers in measuring sights. First he holds his hand, with palm to rear and fingers pointing upward, at such distance from his eye that each finger will measure one sight on the wall. Then, he lowers his hand to his side without changing the angle of the wrist or elbow and notes the exact point at which the hand strikes the body. Thereafter when measuring with the fingers he first places his hand at this point and raises his arm to the front without changing the angle of the wrist or elbow. His hand will then be in the correct position for measuring sights by fingers. The men then determine the proper distance of fingers from the eye as explained by the instructor.

(d) Practice in lateral measurement is given, using convenient objects within view and using both sights and fingers.

c. *Exercise No. 3.*—(1) *Purpose.*—To afford practice in target designation by pointing with the rifle.

(2) *Method.*—(a) The unit or team is formed faced to the rear. The instructor then points out the target to the leader who takes the kneeling or prone position, estimates the range, adjusts his sight, alines his sight to the target, and then calls, "Ready."

(b) The members of the unit or team then move in turn to a position directly behind the leader and look through the sights until they have located the target. The range is given orally by the leader to each individual.

(c) As soon as each man has located the target he moves to the right or left of the leader, sets his sight, places his rifle on a sandbag or other rest, and alines his sights on the target.

(d) The instructor, assisted by the unit or team leader, verifies the sight setting and the alinement of the sights of each rifle.

d. *Exercise No. 4.*—(1) *Purpose.*—To afford practice in target designation by oral description.

(2) *Method.*—(a) The unit or team is deployed faced to the rear. The leader is at the firing point, where sandbags or other rests have been provided for each rifle.

(b) At a prearranged signal the target is indicated by the display of a flag. When the leader states that he understands the position of the target the flag is withdrawn.

(c) The unit or team is then brought to the firing point, placed in the prone position, and each man required to set his sight, use the sandbag or other rest, and sight his rifle on the target according to the oral description of the leader. The leader gives his target designation from the prone position.

(d) The leader's designation is checked from the ground. Then men are required to leave their rifles on the rests, properly pointed, until checked by the instructor or leader.

SECTION IV

RIFLE FIRE AND ITS EFFECT

■ 174. TRAJECTORY.—*a. Nature*.—The trajectory is the path followed by a bullet in its flight through the air. The bullet leaves the rifle at a speed of about 2,700 feet per second. Because of this great speed, the trajectory at short ranges is almost straight or flat.

b. Danger space.—The space between the rifle and the target in which the trajectory does not rise above a man of average height is called the *danger space*. The trajectory for a range of 700 yards does not rise above 68 inches. Therefore, it is said that the danger space for that range is continuous between the muzzle of the gun and the target. For ranges greater than 700 yards, the bullet rises above the height of a man standing, so that only parts of the space between the gun and the target are danger spaces. Figure 61.

■ 175. DISPERSION.—Because of differences in ammunition, aiming, holding, and wind effects, a number of bullets fired from a rifle at a target are subject to slight dispersion. The trajectories of those bullets form an imaginary cone-shaped figure called the *cone of dispersion*.

■ 176. SHOT GROUPS.—When the cone of dispersion strikes a vertical target it forms a pattern called a *vertical shot group*. A shot group formed on a horizontal target is called a *horizontal shot group*. Due to the flatness of the trajectory, horizontal shot groups on level ground vary in length from 100 to 400 yards, depending upon the range.

■ 177. BEATEN ZONE.—The beaten zone is the area on the ground struck by the bullets forming a cone of dispersion. When the ground is level, the beaten zone is also a horizontal shot group. The slope of the ground has great effect on the shape and size of the beaten zone. Rising ground shortens the beaten zone. Ground that slopes downward and in the approximate curve of the trajectories will greatly lengthen the beaten zone. Falling ground with greater slope than the trajectory will escape fire and is said to be in defilade.

174

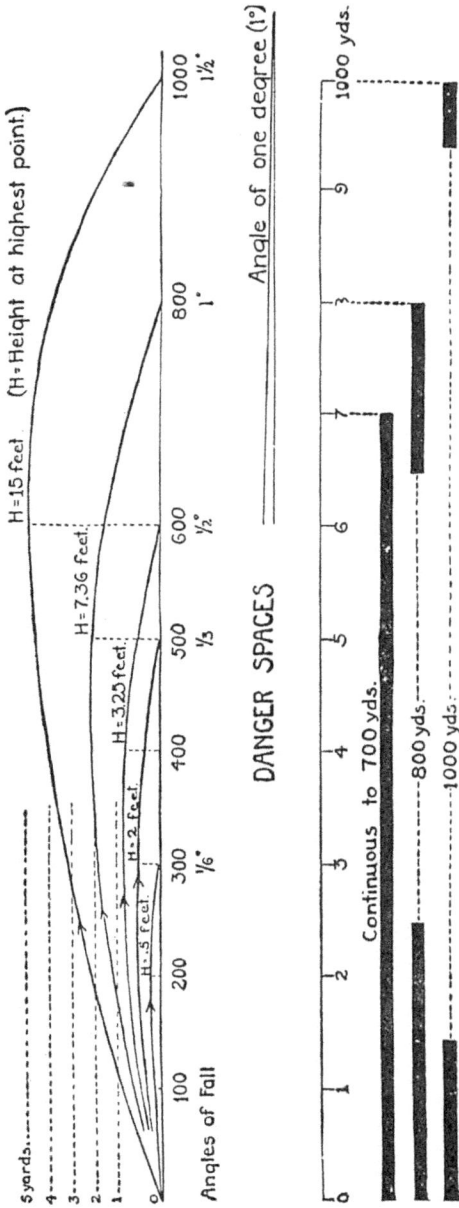

FIGURE 61.—Trajectory diagram (vertical scale is 20 times horizontal scale).

■ 178. CLASSES OF FIRE.—*a*. Fire, as regards direction, is classified as follows:

(1) *Frontal*.—Fire delivered on the enemy from his front.

(2) *Flanking*.—Fire delivered on the enemy from his flank.

b. Fire, as regards trajectory, is classified as follows:

(1) *Grazing*.—Fire approximately parallel to the ground and close enough thereto to strike an object of a given height. The average height of a man (68 in.) is usually taken as determining grazing fire.

(2) *Plunging*.—Plunging fire is fire in which the angle of fall of the bullets with reference to the slope of the ground is such that the danger space is practically confined to the beaten zone and the length of the beaten zone is materially lessened. Fires delivered from high ground on ground lying approximately at right angles to the cone of fire, or against ground rising abruptly to the front with respect to the position of the rifle, are examples of plunging fire. As the range increases, fire becomes increasingly plunging because the angle of fall of the bullets becomes greater.

(3) *Overhead*.—Fire delivered over the heads of friendly troops.

c. *Comparison*.—Flanking fire is more effective than frontal fire. Grazing fire is more effective than plunging fire, because the beaten zone is much longer. Overhead fire with the rifle is unusual and may be employed only when the ground affords protection to the friendly troops.

■ 179. EFFECT OF FIRE.—The fire of automatic riflemen armed with the automatic rifle, caliber .30, M1918, will generally be opened as close to the enemy as possible. Such fire, properly applied, is of decisive effect. It will also be used against low-flying airplanes and against mechanized attacks. The effect of fire on such targets is covered in chapters 3 and 4.

■ 180. DEMONSTRATION OF TRAJECTORIES. — *a*. *Purpose*. — To show trajectories.

b. *Method*.—The unit under instruction watches the firing of a few tracer bullets at targets whose ranges are announced. Ranges of 300, 600, and 800 yards are suitable selections. The flatness of the trajectories is called to the attention of the men.

SECTION V

APPLICATION OF FIRE

■ 181. GENERAL.—*a.* Fire and movement are combined in combat action of units. The application of fire by units is essential to their success.

b. Application of fire in attack.—The automatic rifleman must be trained to place a large volume of accurate fire upon probable enemy locations and indistinct or concealed targets such as enemy machine guns or small groups. The automatic rifleman must be trained to apply such fire quickly upon the order or signal of his leader and in appropriate circumstances to apply it without such order.

c. Application of fire in defense.—In defense the fire of automatic riflemen is delivered from positions which must be held. They are placed to secure good field of fire covering probable avenues of approach either on the ground or in the air, and to take advantage of cover and concealment.

■ 182. CONCENTRATED AND DISTRIBUTED FIRE.—The size and nature of the target presented may call for the fire power of the entire unit or team or only certain parts. The fire of a unit or team must necessarily be either concentrated or distributed fire.

a. Concentrated fire.—Concentrated fire is fire directed at a single point. This fire has great effect but only at a single point. Antitank guns and automatic weapons are examples of suitable targets for concentrated fire.

b. Distributed fire.—(1) Distributed fire is fire distributed in width for the purpose of keeping all parts of the target under effective fire. It is habitually used on targets having any considerable width such as a portion of the edge of a woods or road.

(2) Unless otherwise instructed the automatic rifleman will habitually cover the entire target. The first shot is fired on that portion of the target corresponding generally to the automatic rifleman's position in the unit or team.

(3) If other targets appear, the unit or team leader announces such changes in the fire distribution as are necessary.

177

■ 183. ASSAULT FIRE.—Assault fire is that automatic fire delivered by the automatic rifleman while steadily advancing at a walk. Automatic riflemen load while advancing, the rifle being carried as prescribed in paragraph 66.

■ 184. RATE OF FIRE.—The automatic rifleman fires at the rate of fire most effective under existing conditions and generally at a rate of from 40 to 60 shots per minute semiautomatic fire. An excessive rate wastes ammunition without corresponding effect.

■ 185. FIRE DISCIPLINE.—Fire discipline in the unit or team includes the careful observance of the instructions relative to the use of the automatic rifle in combat and exact execution of the orders of the leader. It implies care in sight setting, aim, trigger squeeze, close attention to the leader, independent increase in the rate of fire when the target becomes more favorable, cessation of fire on the unit or team leader's order or signal, or when a target cannot be located with sufficient definition to justify the expenditure of ammunition. It also implies that when the unit or team leader has released the automatic riflemen from the control of his fire order, the automatic rifleman acts on his own initiative, selects sight setting and target independently, and opens and ceases fire in accordance with the situation.

■ 186. FIRE CONTROL.—*a.* Fire control consists of the initiation and supervision of the fire of the unit or team by its leader. By initiating fire on order or signal the effect of surprise is increased. On the other hand the irregular formations adopted for an advance will often render such action impracticable. In such case fire must be opened and maintained on the initiative of individuals as circumstances require. In any case the leader of the unit or team must supervise and seek to control the fire of his men so that it is directed and maintained at suitable targets. All must understand that controlled fire is always the most effective.

b. How exercised.—Unit or team leaders, assisted by their seconds-in-command, exercise fire control by means of orders, commands, and signals. The signals most frequently used are—

SIGNALS FOR RANGE.
COMMENCE FIRING.

FIRE FASTER.

FIRE SLOWER.

CEASE FIRING.

ARE YOU READY?

I AM READY.

A description of the above signals is found in FM 22–5 and FM 2–5.

■ 187. FIRE ORDERS.—*a. Purpose.*—The leader of a unit or group, having made a decision to fire on a target, must give certain instructions as to how the target is to be engaged. The instructions by which the fire of a unit or team is directed and controlled form the fire order.

b. Basic elements of a fire order.—A fire order contains three basic elements, which are announced or implied in every case. Only such elements or parts thereof will be included as are essential. The sequence is always as follows:

Target-designation element.

Fire-distribution element.

Fire-control element.

(1) *Target-designation element.*—The target may be designated by any one, or a combination, of the prescribed methods. (See sec. III, ch. 5.)

(2) *Fire-distribution element.*—The fire-distribution element is normally omitted from the fire order for rifle units. The method of fire distribution described in paragraph 182b is habitually employed. When necessary, the fire-distribution element includes the subdivision of the target. For example—

(*a*) A leader desires to engage two machine-gun nests; the distribution element of his order might be as indicated by the italicized words below.

Range: 500.

Front.

Machine gun at base of lone pine.

Cooper, or *Cooper, Emerson, Crane, Hines, Jones.*

Range: 500.

Left flank.

Machine gun at base of haystack.

Brown, or *Brown, Smith, Turner, Howard, Stone.*

(*b*) The unit or team leader may engage several targets by placing one automatic rifleman or one automatic rifleman and several riflemen under the command of an assistant and directing him to engage one target, while he engages another with the remainder of the unit or team.

(3) *Fire-control element.*—The fire-control element normally consists initially of merely the command or signal COMMENCE FIRING. It may include the number of magazines or rounds. Other fire-control elements are—

AT MY SIGNAL (followed by hand signal).

ONE MAGAZINE (FIVE ROUNDS) COMMENCE FIRING.

(4) Example of a complete fire order follows:

(a) *Target-designation element.*

(Range) _____Range: 600.

(Direction) _____ Reference: right edge of lone building.

(Description of target) __Target: group of enemy along hedge.

(b) *Fire-distribution element.*—(Implied.)

(c) *Fire-control element.*—COMMENCE FIRING.

■ 188. DUTIES OF LEADERS.—The following summary of duties of leaders relates only to their duties in the technique of fire.

a. Unit or team leader.—(1) Carries out orders of higher commanders.

(2) Selects firing positions for unit or team.

(3) Designates targets and issues fire orders.

(4) Controls fire of unit or team.

(5) Maintains fire discipline.

(6) Observes targets and effect of fire.

b. Second-in-command.—(1) Carries out orders of unit or team leader.

(2) Assists the leader to maintain fire discipline.

(3) Assumes command of unit or team in absence of leader.

(4) Participates in firing when the fire of his rifle is considered more important than other assistance to the unit or team leader.

SECTION VI

LANDSCAPE-TARGET FIRING

■ 189. SCOPE AND IMPORTANCE.—*a.* After satisfactory progress has been made in the preceding steps, the automatic rifleman may be given practice in the application of those lessons by firing at landscape targets.

b. The advantages of this training are as follows:

(1) Close supervision over all members of the firing group is made possible by their close proximity.

(2) Accessibility and nature of the targets permit the application and effect of the fire to be readily shown.

(3) It is a form of instruction which lends itself to indoor training when lack of facilities or weather conditions make it desirable or necessary.

c. In circumstances where there is a choice between landscape-target firing as covered in this section and firing at field targets as covered in section VII, the latter is to be preferred.

■ 190. DESCRIPTION OF TARGET.—A landscape target is a panoramic picture of a landscape, and is of such size that all, or nearly all, of the salient features will be recognizable at a distance of 1,000 inches. The standard target is the series A target of five sheets in black and white.

■ 191. WEAPONS TO BE USED.—Firing at landscape targets should be done with caliber .22 rifles, preferably the M1922M2 equipped with the Lyman receiver sight. When a sufficient number of those rifles are not available, the Browning automatic rifle, caliber .30, M1918, may be used.

■ 192. PREPARATION OF TARGETS.—*a. Mounting.*—(1) The sheets are mounted on frames made of 1- by 2-inch dressed lumber, with knee braces at the corners. The frames for the target sheets are 24 by 60 inches. These frames are covered with target cloth which is tacked to the edges.

(2) The target sheets are mounted as follows: Dampen the cloth with a thin coat of flour paste and let it dry for about an hour; apply a coat of paste similarly to the back of the paper sheet and let it dry about an hour; apply a second coat of paste to the back of the paper and mount it on the cloth; smooth out wrinkles, using a wet brush or sponge, and work from the center to the edges. The frame must be placed on some surface which will prevent the cloth from sagging when the paper is pressed on it. A form for this purpose can easily be constructed. It must

be of the same thickness as the lumber from which the frames are built, and must have approximately the same dimensions as the aperture of the target frame.

b. Supports for target frames.—The target frames described above are set on posts placed upright in the ground, 5 feet from center to center. The target frames are supported on the posts by cleats and dowels in order to allow for easy removal.

c. Range indicators.—In order to make all elements of target designation complete, assumed ranges must be used on landscape targets. Small cards on which are painted appropriate numbers representing yards of range are tacked along one or both edges of a series of panels. The firers must be cautioned that the range announced in any target designation is for the sole purpose of designating the target, and that the sight setting necessary to zero their rifles must not be changed.

d. Direction cards.—In order to provide the direction element in oral target designation, small cards on which are painted Front, Right front, Left front, Right flank, Left flank are tacked above the appropriate panels of the landscape series.

e. Scoring devices.—(1) A unit or team may be brought up to the target and there view the results of its firing. Scoring the exercises will tend to create competition between units or teams and will enable the instructor to grade their relative proficiency in this form of training. A scoring device, conforming in size to the 50 and 75 percent shot groups to be expected of average shots firing at 1,000 inches, and at reduced ranges, can easily be made from wire, or a better one may be prepared by imprinting a scoring diagram on a sheet of transparent celluloid. The scoring space is outlined on the target in pencil before the target is shown to unit or team leaders. This procedure prevents any misunderstanding of leaders as to the limits of the designated target. Upon completion of firing the entire unit or team is shown the target and the results of the firing.

(2) While shot groups are in the form of a vertical ellipse, the 50 and 75 percent zones should be shown by the devices as rectangles. This is for convenience in their preparation.

For a distance of 1,000 inches the 50 percent zone is a rectangle 2½ inches high by 2 inches wide; the 75 percent rectangle is 5 inches high by 4 inches wide. For a distance of 50 feet the 50 percent zone is a rectangle 1½ inches high by 1.2 inches wide; the 75 percent rectangle is 3 inches high by 2.4 inches wide. The target is at the center of the inner rectangle, or 50 percent zone.

(3) For a linear target, such as a small area over which the automatic riflemen will distribute their fire, the 50 percent zone is formed by two parallel lines, drawn parallel to the longer axis of the target (area) and with the target midway between those lines. For a distance of 1,000 inches the lines should be 2½ inches apart; for a distance of 50 feet the lines should be 1½ inches apart. Two additional lines, similarly drawn, form the 75 percent zone. For a distance of 1,000 inches the lines should be 5 inches apart; for a distance of 50 feet the lines should be 3 inches apart. The width of the zones will vary according to the size of the target selected. For a distance of 1,000 inches, the zones extend 1 inch beyond each end of the target; for a distance of 50 feet the zones extend 0.6 inch beyond each end of the target. The zones are then divided into a convenient number of equal parts, the number depending on the length (width) of the target and the number of men firing. This is done in order to give a score for distribution of shots fired on a linear target (see par. 195b).

■ 193. ZEROING-IN OF RIFLES.—a. It will be necessary to zero-in the rifles used before firing exercises on the landscape target. A blank target with a row of ten 1-inch-square black pasters about 6 inches from and parallel with the bottom edge of the target should be prepared and used for this purpose. In all firing for zeroing-in, sandbag rests are used.

b. The procedure in detail is as follows:

(1) The sights of the rifle are blackened.

(2) The squad is deployed on the firing points; the leader takes the proper position in rear of the squad.

(3) The instructor causes each firer to set his sights at zero elevation and zero windage, and checks each sight.

(4) Each man is assigned the particular small black paster which corresponds to his position in the squad as his aiming point.

(5) Three rounds are issued to each man on the firing point and are to be loaded and fired singly at the command of the instructor.

(6) Each man fires three shots at his spotter at the command THREE ROUNDS, COMMENCE FIRING.

(7) The instructor commands: CLEAR RIFLES. The unit or team leader checks to see that this is done.

(8) The instructor and unit or team leader inspect the target and, based upon the location of the center of impact of the resultant shot group, gives each man the necessary correction for his next shot, as "Up 1 minute, right one-half point."

(9) The firing continues as outlined above until all rifles are zeroed-in, that is, until each man has hit his aiming point.

c. For a caliber .22 rifle with the Lyman receiver sight, at a distance of 1,000 inches, a change of 5 minutes in elevation will move the strike of the bullet about 1½ inches. A change of one point of windage moves the strike about 1¼ inches. At a distance of 50 feet a change of 6 minutes in elevation will move the strike of the bullet about 1 inch, and a change of one point of windage, about ¾ of an inch.

d. To zero the Browning automatic rifle, caliber .30, M1918, at 1,000 inches, see paragraph 85.

■ 194. FIRING PROCEDURE.—The following is the sequence of events in conducting firing exercises:

a. All members of the unit or team, except the leader, face to the rear.

b. The instructor takes the leader to the panels and points out the target to him.

c. They return to the firing point; the leader takes charge of the unit or team and causes the men to resume their firing positions.

d. The leader gives the command LOAD, cautioning, "—— rounds per rifleman and —— rounds per automatic rifleman only."

e. The leader designates the target orally. Reference to panels to indicate direction should not be allowed in the designation. To complete the fire order, the leader adds, COMMENCE FIRING.

f. When the unit or team has completed firing, the leader commands: CEASE FIRING, CLEAR RIFLES. The unit or team then examines the target. The target panel is scored and marked with the unit or team number.

g. The instructor holds a short critique after each exercise.

■ 195. SCORING.—*a. Concentrated fire.*—In concentrated fire the sum of the value of the hits within the two zones is the score for the exercise. For convenience of scoring and comparison, 100 is fixed as the maximum score. Any method of scoring and of distribution of ammunition among members of the unit or team may be used. The following examples based on firing 50 rounds are given as suggested methods:

(1) Value of each hit in 50 percent zone, 2.

(2) Value of each hit in 75 percent zone, 1.

b. Distributed fire.—The following is a method of scoring for distributed fire of the unit or team on a target of width.

(1) Value of each hit in 50 percent zone, 2.

(2) Value of each hit in 75 percent zone, 1.

(3) Value of each distribution space (if target is divided into 10 equal spaces), 10.

(4) The score for distribution, plus the value of all hits, divided by two is the score for the exercise.

■ 196. EXERCISES.—The fire of the automatic rifleman armed with the caliber .22 rifle will be employed against targets appropriate to automatic rifle fire and as though such weapons were being fired.

a. Exercise No. 1.—(1) *Purpose.*—To teach target designation and to show the effect of concentrated fire.

(2) *Method.*—The unit or team leader employs the fire of his unit or team at one point-target indicated to him by the instructor.

b. Exercise No. 2.—(1) *Purpose.*—To teach target designation and the division of fire on two points of concentration.

(2) *Method.*—The instructor indicates two point-targets to the unit or team leader, giving the nature of each. The unit or team leader applies the fire of part of his unit or team on

one target and the fire of the remainder of his unit or team on the other. The scoring will be as for concentrated fire on each target, the several scores being combined in totals for the score for the exercise.

c. *Exercise No. 3.*—(1) *Purpose.*—To teach target designation and fire control in diverting part of the fire of the unit or team to a suddenly appearing target.

(2) *Method.*—The instructor indicates a point-target to the unit or team leader. The leader applies the fire of his unit or team to the target. After firing has commenced, the instructor indicates and gives the nature of a new target to a flank. When the second target is indicated, the leader shifts the fire of part of his unit or team, as directed by the instructor, from the first to the second target.

d. *Exercise No. 4.*—(1) *Purpose.*—To teach target designation, fire control, and the method of searching a small area with automatic rifle fire.

(2) *Method.*—The instructor indicates and gives the nature of two point-targets. The leader applies the fire of his unit or team on the two point-targets as directed by the instructor. After firing has commenced, the instructor indicates a small area in which an enemy group is under cover. When the area target is indicated, the leader is told to shift the fire of an automatic rifleman to that target.

e. *Exercise No. 5.*—(1) *Purpose.*—To teach the application of fire on an enemy group marching in formation, the fire control necessary to obtain fire for surprise effect, and to show the effect of fire on troops in formation.

(2) *Method.*—The instructor indicates to the leader a target that represents a small group of the enemy marching in approach march formation, formation for patrol, or the like, the enemy not being aware of the presence of the unit or team. The leader applies the fire of his unit or team; his instructions must result in the simultaneous opening of fire of all weapons and the distribution of fire over the entire target. The assignment of half of his unit or team to fire at the rear half of the target and the remainder of his unit or team at the forward half is a satisfactory method of distributing fire over such target.

Section VII

FIRING AT FIELD TARGETS

■ **197. Scope of Training.**—The training in this step is similar to that given the soldier in landscape-target firing, but with the added feature of firing the Browning automatic rifle, caliber .30, M1918, at field targets at unknown ranges, the use of cover, fire control under more usual conditions, and range estimation. In order to make this training progressive, the automatic rifleman is first given an opportunity to fire at partially exposed field targets of unknown ranges. As a final stage in this instruction he will be required to fire at some targets which are concealed from view but exposed to fire. Individuals preferably receive this training in the squad or in smaller groups.

■ **198. General Considerations.**—*a. Progressive training.*—The inclusion of the training in moving from an approach march formation, or place of concealment, to firing positions is primarily to teach the soldier the proper use of cover and selection of firing positions and to connect up the technique of applying and controlling collective fire with other prerequisite applied subjects.

b. Firing positions and representation of enemy.—In battle a unit is not deployed with individuals abreast and at regular intervals apart. The selection of individual and group positions is governed by the field of fire, cover or concealment while firing, cover of approach to those positions, fire control, and nature of target. The representation of the enemy will conform to irregular battle formations. Safety precautions necessary in firing at field targets are given in paragraph 199.

c. Use of cover.—(1) The individual use of cover and concealment is taught in FM 21–45 (now published as ch. 9, BFM vol. I). In training in firing at field targets the principles are the same.

(2) In seeking cover in a firing position men may move a few yards in any direction, but they must not be allowed to bunch together behind concealment which does not afford protection from fire. They avoid positions which will mask

the fire of others or cause their own fire to be dangerous to other men of their unit.

d. Marksmanship applied. — (1) The fundamentals of known-distance automatic rifle marksmanship are followed in this training insofar as they are applicable to field conditions.

(2) The fundamentals of known-distance marksmanship should be applied to the technique of fire and to combat in a common-sense way. For example, it will often be impracticable to keep the sights blackened and the soldier is permitted to take advantage of trees, rocks, or any other rest for his weapon which will make his fire more accurate.

e. Use of battle sight.—The battle sight corresponds to a sight setting of approximately 550 yards. It is used on targets from 0 to 600 yards when time is lacking for setting the sight, or in firing at moving targets. By keeping the sight habitually set at 300 yards when not in use, the soldier has two sights set ready for emergencies.

■ 199. SAFETY PRECAUTIONS.—*a.* In general, the safety precautions used on the known-distance ranges apply with equal force to instruction in the technique of fire. Safety of personnel is of primary importance in conducting exercises which require the firing of service ammunition. To this end exercises should be drawn to conform to the state of training of the units concerned.

b. The officer in charge of an exercise is responsible for the safety of the firing; it is his duty to initiate and enforce such precautions as he deems necessary under existing conditions. No other officer can modify his instructions without assuming the responsibility for the safety of the firing.

c. Before permitting fire to be opened all men should be on a general line. No man should be permitted to be ahead of or in rear of this line, a distance greater than one-half the interval between himself and the man next to him. For example, if the interval between men is 10 paces, then no man should be more than 5 paces ahead of or behind the man next to him.

d. Ball ammunition will not be loaded until each man is in the firing position and the officer in charge has insured that it is safe for each man to fire. Upon completion of firing the officer in charge will cause all rifles to be unloaded and inspected, and the ammunition to be collected.

e. Upon completion of the day's firing automatic rifles and belts will be inspected by an officer to insure that no ammunition remains in them.

f. Special precautions will be taken to insure that the range is clear before ammunition is issued.

g. During the firing of exercises rifles will be pointed in the direction of the targets at all times. Special vigilance is required to enforce this rule while men are using a cleaning rod to remove an obstruction from the chamber.

■ 200. SITUATIONS FOR FIRING EXERCISES.—*a.* Each exercise should be initiated by a unit—

(1) Already deployed in a firing position.

(2) Halted in approach march formation or in a place of concealment with observers out.

b. In the first case, each man should be in a selected firing position, special attention being paid to individual cover and concealment.

c. In the second case unit or team leaders select the firing positions for the members of their units or teams; they conduct their units or teams forward by concealed routes and send the automatic riflemen to their firing positions by individual directions. Occupation of the initial firing position of a unit is done with the minimum of exposure.

■ 201. CRITIQUE.—At the completion of the firing of any exercise the instructor should conduct a critique of that exercise with the firing unit. A suggested form for such a critique is as follows:

a. Purpose of the exercise.

b. Approach and occupation of the firing position (individual concealment and cover).

c. Fire order (particular reference being made to the target-designation element).

d. Time required to open fire (from the time the leader is told the range is clear).

e. Rate of fire.

f. Fire control.

g. Effect of fire. (Upon completion of firing and when the range is clear, the targets are scored.)

h. Performance of the unit satisfactory or unsatisfactory.

■ 202. SUGGESTED EXERCISES.—*a. Exercise No. 1.*—(1) *Purpose.*—Practice in fire orders, application of the fire of a unit or team in position, fire control, and proper individual concealment in the occupation of the firing position.

(2) *Method.*—Enemy is represented by one group of targets exposed to fire but partially concealed from view, requiring a simple fire order. The unit or team leader is shown the targets (personnel with flag) and safety limits for firing position of the unit or team. When the unit or team leader fully understands the location and nature of the target and the instructor informs him that the range is clear, he will load ball ammunition, give the fire order, and fire the problem. The range should be estimated by eye and the target designated by oral description.

b. Exercise No. 2.—(1) *Purpose.*—Practice in fire orders, application of the fire of a unit or team armed with the Browning automatic rifle, caliber .30, M1918, on a linear target, fire control, proper deployment and individual concealment in the occupation of the firing position, and engagement of a surprise target.

(2) *Method.*—Silhouette targets, representing an enemy unit or team deployed in a firing position, are partially concealed from view but exposed to fire. A screen behind the targets is marked with distribution spaces to give unit or team credit for the shots that did not hit the targets but which would have had an effect on an enemy. The unit or team is in rear of the firing position; unit or team leader is shown the linear target (by flag) and then conducts unit or team forward and disposes it in a concealed firing position. When unit or team leader is told the range is clear, he will engage the target with surprise fire. A *surprise target,* well to the flank of the first target, representing an enemy machine gun, appears shortly after the unit or team has engaged the linear target. The unit or team leader is told the amount of fire to shift to the surprise target. In addition to the suggested form of critique in paragraph 201, proper distribution of fire on a linear

target and the engagement of the surprise target should be discussed.

c. Exercise No. 3.—(1) *Purpose.*—Practice in the application of automatic rifle fire over a small area in which an enemy is concealed.

(2) *Method.*—Targets are placed within a small area, exposed to fire but concealed from view. An automatic rifleman is directed to search that area with fire. He distributes his fire throughout the length and breadth of the area, using a rapid rate of fire.

d. Exercise No. 4.—(1) *Purpose.*—Practice in firing at moving targets.

(2) *Method.*—Automatic riflemen fire individually at targets carried on long sticks by men in the pits of a class A range. The men in the pits are each assigned a space, the width of about five regular range-target spaces, in which they walk continuously back and forth. By whistle signal, targets are exposed to the firing line for 5 seconds and then concealed for 5 seconds. Targets are exposed once for each shot to be fired. On the firing line one man is assigned to each target. Ranges of 200 or 300 yards are best suited for this class of firing.

CHAPTER 6

ADVICE TO INSTRUCTORS

                                                              Paragraphs
SECTION   I. General_____   203
         II. Mechanical training_____   204
        III. Marksmanship—known-distance targets_____ 205–219
         IV. Marksmanship—air targets _____ 220–224
          V. Technique of fire _____ 225–233

SECTION I

GENERAL

■ 203. PURPOSE.—The provisions of this chapter are to be accepted as a guide and will not be considered as having the force of regulations. They are particularly applicable to emergency conditions when large bodies of troops are being trained under officers and noncommissioned officers who are not thoroughly familiar with approved training methods.

SECTION II

MECHANICAL TRAINING

■ 204. POSITION STOPPAGE SET-UPS.—*a. First-position stoppages.*—(1) Place blown primer between lips of magazine and top cartridge. Let bolt go forward. Replace magazine.
Answer: Failure to feed—change magazine.

(2) Place empty cartridge case in chamber. Let bolt go forward. Replace magazine.
Answer: Insufficient gas—correct gas adjustment.

b. Second-position stoppage.—Place blown primer on face of bolt or up in locking recess. Let bolt go forward. Replace magazine.
Answer: Obstruction—remove blown primer.

c. Third-position stoppage.—Cock rifle, then place ruptured case in chamber. Replace magazine. Let bolt go forward.
Answer: Call for ruptured-cartridge extractor.

d. Fourth-position stoppage.—(1) Cock rifle. Insert empty cartridge case in chamber. Replace loaded magazine. Have man pull trigger.
Answer: Call for cleaning rod—examine extractor.

192

(2) Set change lever on "safe."

(3) Remove connector from trigger guard or have middle prong of sear spring resting on wall of sear carrier.

SECTION III

MARKSMANSHIP—KNOWN-DISTANCE TARGETS

■ 205. GENERAL.—*a.* Training is preferably organized and conducted as outlined in paragraphs 55 and 56. Officers should generally be considered as the instructors of their units. As only one step is taken up at a time, and as each step begins with a lecture and a demonstration showing exactly what to do, the trainees, although not previously instructed, can carry on the work under the supervision of the instructor.

b. It is advisable that personnel to fire be relieved from routine garrison duty during the period of preparatory marksmanship training and range practice with the automatic rifle.

■ 206. PLACE OF ASSEMBLY FOR LECTURES.—Any small ravine or cup-shaped area makes a good amphitheater for giving the lecture in case no suitable building is available.

■ 207. ASSISTANT INSTRUCTORS.—*a.* It is advantageous to have all officers and as many noncommissioned officers as possible trained in advance in the prescribed methods of instruction. When units are undergoing automatic rifle marksmanship training for the first time, this is not always practicable. A good instructor can give a clear idea of how to carry on the work in his lecture and demonstration preceding each step. In the supervision of the work following the demonstration, he can correct any mistaken ideas or misinterpretations.

b. When an officer in charge of automatic rifle instruction is conducting successive organizations through target practice, it is advisable to attach officers and noncommissioned officers of the units to follow to the first organization taking the course for the period of preparatory work and range firing. These act as assistant instructors when their own companies take up the work. Such assistants are particularly useful when one group is firing on the range and another is going through the preparatory exercises, both under the supervision of one instructor.

■ 208. EQUIPMENT.—*a.* All equipment used in the preparatory exercises must be accurate and carefully made. One of the objects of these exercises is to cultivate a sense of exactness in the minds of the men undergoing instruction. They cannot be exact with poor equipment.

b. The instructor should personally inspect the equipment for the preparatory exercises before the training begins. A set of model equipment should be prepared in advance by the instructor for the information and guidance of the organization about to take up the preparatory work. The sighting bars must be made as described, and the hole representing the peep sight must be absolutely circular. If the sights are made of tin, the holes should be bored by a drill. Good rear sights can be made for the sighting bars by using cardboard and cutting the holes with a punch of the type used for cutting wads for 10-gage shotgun shell. Silhouettes painted on a white background are not satisfactory. The silhouette target from the M1 1,000-inch target pasted on tin or stiff backing makes the best aiming points either for sighting and aiming exercises or for use in position and trigger-squeeze exercises.

■ 209. INSPECTION OF RIFLES.—No man is required to fire with an unserviceable or inaccurate rifle. All automatic rifles should be carefully inspected far enough in advance of the period of training to permit organization commanders to replace all inaccurate or defective rifles. Rifles having badly pitted barrels are not accurate and should not be used.

■ 210. AMMUNITION.—The best ammunition available should be reserved for record firing and the men should have a chance to learn their sight settings with that ammunition before record practice begins. Ammunition of different makes and of different lots should be used indiscriminately.

■ 211. ORGANIZATION OF WORK.—*a. In preparatory training.*— (1) The field upon which the preparatory work is to be given should be selected in advance and a section of it assigned to each group. The equipment and apparatus for the work should be on the ground and in place before the morning lecture is given, so that each group can move to its place and begin work immediately and without confusion. Figure 62 shows a suggested organization for the work when a number

200 yd aiming targets

Aiming boxes ➞

Group   Group

Rifle Rests ➞

Center line. One-half
company on each side

Group   Group

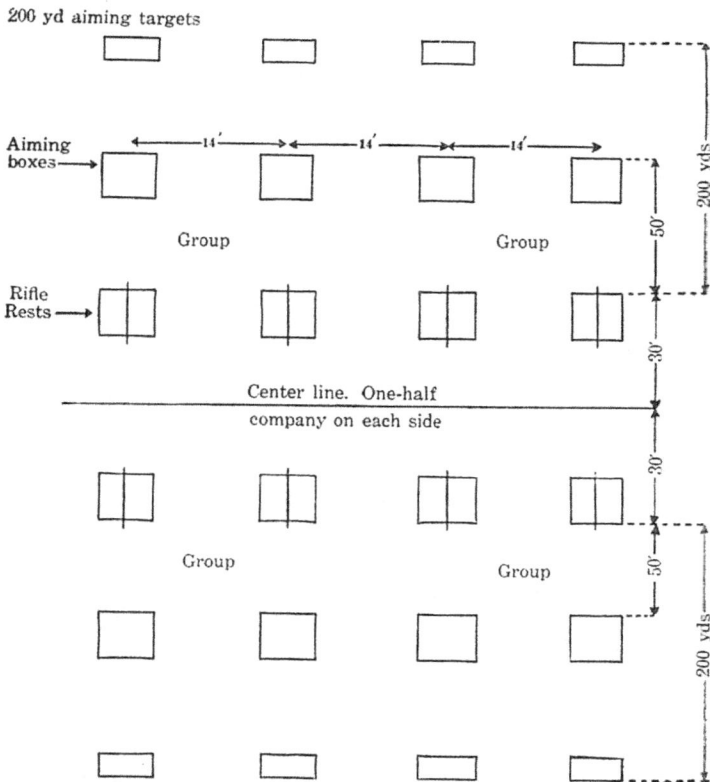

FIGURE 62.—Portion of field laid out for sighting and aiming
exercises.

of groups are undergoing instruction at the same time. In
this way the instructors, whose positions are normally between
the lines, have all of their men under close supervision.

(2) The arrangement of the equipment is as follows:

(a) On each line are placed the sighting bars and rifle rests
at sufficient intervals to permit efficient work.

(b) Fifty feet from each line is placed a line of small boxes
with blank paper tacked on one side, one box and one small
sighting disk to each rifle rest.

(c) Two hundred yards from each line is placed a line of
frames suitable for use in making shot groups at 200 yards,
one frame to each unit the size of a squad. These frames

have blank paper tacked or pasted on the front. A long-range sighting disk is placed with each frame. Machine-gun targets make acceptable frames for this work.

(3) In position and trigger-squeeze exercises, targets should be placed at 1,000 inches and 200 yards.

(4) When sufficient level ground is not available for the above arrangement, the organizations will have to vary from it in some particulars. It will nearly always be found, however, that all of the work except making shot groups at 200 yards can be carried on in two lines.

b. In range practice.—(1) The range work should be so organized that there is a minimum of lost time on the part of each man. Long periods of inactivity while awaiting a turn on the firing line should be avoided. For this reason the number of men on the range should be accommodated to the number of targets available.

(2) As a general rule six men per target is about the maximum and four men per target the minimum for efficient handling.

■ 212. MODEL SCHEDULES.—The following schedules are suggested for guides in a course in preparatory marksmanship and firing course A:

a. Preparatory training.—(1) *First day.*

Subject	Time allotted (hours)
Purpose of preparatory marksmanship training	¼
First step—sighting and aiming exercises:	
Explanation and demonstration	½
First sighting and aiming exercise	½
Sight blackening and second sighting and aiming exercise	¾
Third sighting and aiming exercise	2
Second step—position exercises:	
Explanation and demonstration	½
Gun sling adjustment; trigger slack; holding the breath; general rules for positions	½
Prone position, including sandbag rest	1
Sitting position	½
Kneeling position	1
Assault fire	½

(2) *Second day.*

Subject	Time allotted (hours)
Review of positions	1½
Third step—trigger-squeeze exercises:	
Explanation and demonstration	½
Trigger-squeeze exercise, prone position (sandbag optional)	½
Trigger-squeeze exercise, sitting	½
Trigger-squeeze exercise, kneeling	1
Trigger-squeeze exercise, prone	1
Effect of wind: sight changes; use of score book	1
Examination of all men by section and platoon leaders in all preparatory subjects and exercises [1]	2

[1] Lack of proficiency disclosed by examination will be corrected at once by additional instruction.

NOTE.—The use of the scorebook and effects of light and wind will be taken up with men who are not actually on the line undergoing instruction.

b. Range practice, course A.—(1) *Third day.*

Subject	Time allotted (hours)
Rapid fire exercises, 1,000-inch range. Fire tables I and II (each score preceded by a simulated run for each man)	8

(2) *Fourth day.*

Subject	Time allotted (hours)
Fire tables III and IV (each score preceded by a simulated run)	8

(3) *Fifth day.*

Subject	Time allotted (hours)
Fire table V (each score preceded by a simulated run)	4
Fire table VI (each score preceded by a simulated run)	4

c. *Record practice, course A.—Sixth day.*

Subject	Time allotted (hours)
Fire table VII	8

NOTE.—The time allotted for firing the known-distance range is based on six orders per target and a simulated run preceding each practice for each man.

d. *Courses B, C, and D.*—The preparatory exercises and 1,000-inch firing are the same as course A. All other firing is conducted in a manner similar to course A, reducing the time accordingly.

■ 213. LECTURES AND DEMONSTRATIONS.—*a.* The lecture at the beginning of each step are an important part of the instructional methods. The lectures may be given to the assembled groups undergoing preparatory automatic rifle training up to and including all the automatic riflemen of a regiment. However, when a battalion takes up automatic rifle training the talks and demonstrations as a rule are made by an officer of each company. It is not necessary that they be expert shots.

b. The notes on lectures which follow are to be used merely as a guide. The points which experience has shown to be the ones which usually require elucidation and demonstration are placed in headings in italics. The notes which follow each heading are merely to assist the instructor in preparing his lectures. The lecturer should know in advance what he is going to say on the subject. Under no circumstances will he read over to a class the outlines for lectures contained herein, nor will he read a lecture prepared by himself. During the lecture the headlines in italics made out by himself serve as a guide as to the order in which the subjects are to be discussed. If he cannot talk interestingly and instructively on each subject without further reference to notes, he should not give the lectures at all.

c. It is important to show the men undergoing instruction, by explanation and demonstration, just how to go through the exercises and to tell them why they are given these exercises.

■ 214. FIRST LECTURE: SIGHTING AND AIMING.—*a.* The class is assembled in a building or natural amphitheater in the open, where all can hear the instructor and see the demonstrations.

b. The following equipment is necessary for the demonstrations:

(1) One sighting bar.

(2) One automatic rifle rest.

(3) One automatic rifle.

(4) One small sighting disk.

(5) One long-range sighting disk.

(6) One small box.

(7) Material for blackening sights.

c. The following subjects are the ones usually discussed in the first lecture.

(1) *Value of knowing how to shoot.*—(*a*) Expertness in the use of the automatic rifle gives the individual confidence and a higher morale.

(*b*) Individual proficiency increases the efficiency of infantry as a whole.

(*c*) Automatic rifle firing is good sport.

(2) *Object of target practice.*—(*a*) To teach men how to shoot.

(*b*) To show them how to teach others.

(*c*) To train future instructors.

(3) *Training to shoot well.*—(*a*) Any man can be taught to shoot well. Shooting is a purely mechanical operation which can be taught to anyone physically fit to be a soldier.

(*b*) It requires no inborn talent such as to play a violin or paint a picture.

(c) There are only a few simple things to do to shoot well, but these things must be done in a manner exactly right. If they are done in a manner only approximately right, the results will be poor.

(4) *Method of instruction.*—(a) The method of instruction is the same as in teaching any mechanical operation.

(b) The instruction is divided into steps. The man is taught each step and practices it before going to the next step. When he has been taught all of the steps he is taken to the rifle range to apply what he has learned.

(c) If he has been properly taught the various preparatory steps he will do good shooting from the very beginning of range practice.

(d) Explain coach-and-pupil method. Why used.

(5) *Reflecting attitude of instructor.*—If the instructor is interested, enthusiastic, and energetic, the men will be the same. If the instructor (unit or team leader) is inattentive, careless, and bored, the men will be the same and the scores will be low.

(6) *Examination of men on preparatory work.*—Each man is examined in the preparatory work before going to the range. An outline of this examination is given in paragraph 75.

(7) *Method of marking blank form.*—Explain blank form, paragraph 75. Explain marking system by the use of a blackboard, if available.

(8) *Five essentials to good shooting.*—(a) Correct sighting and aiming.

(b) Correct position.

(c) Correct trigger squeeze.

(d) Correct application of rapid-fire principles.

(e) Knowledge of proper sight adjustments.

(9) *Today's work.*—First step, sighting and aiming.

(10) *Demonstration of first sighting and aiming exercise.*—Have a unit or team on stage or platform and show just how this exercise is carried on.

(11) *Blackening sights.*—Explain why and demonstrate how this is done.

(12) *Demonstration of second sighting and aiming exercise.*—Assume that some of the squad have qualified in the

first exercise. Put these men through the second sighting and aiming exercise and show just how it is done.

(13) *Demonstration of third sighting and aiming exercise.—*(a) Assume that some of the squad have qualified in the second sighting and aiming exercise. Put these men through the third sighting and aiming exercise and show just how it is done.

(b) Show how the unit or team is organized by the coach-and-pupil method so as to keep each man busy all the time.

(14) *Long-range shot group work.—*Show the class the disk for 200-yard shot group work. Explain how this work is carried on and why. Show some simple system of signals that may be used.

(15) *Final word.—*(a) Start keeping your blank form today.

(b) Organize your work so that all men are busy at all times.

(16) Are there any questions?

(17) Next lecture will be _____. (State hour and place.)

■ 215. SECOND LECTURE: POSITION.—*a.* The following equipment is necessary for the demonstrations in this lecture:

(1) One automatic rifle with sling.

(2) One sandbag.

(3) One box with small aiming target.

b. The following subjects are the ones usually discussed in the second lecture:

(1) *Importance of each step.—*(a) Each step includes all that has preceded.

(b) Each step must be thoroughly learned and practiced or the instruction will not be a success.

(2) *Necessity for correct positions.—*No excellent shot varies from the normal positions. Few men with poor positions are even fair shots. Few men with good positions are poor shots. Instruction in positions involves correct aiming.

(3) *Gun sling.—*Demonstrate both of the gun sling adjustments and explain why they are used and when each is used.

(4) *Taking up slack.—*Show the class the slack on the trigger. Explain why it is taken up in the position exercises.

(Cannot begin to press the trigger until the slack has been taken up.)

(5) *Holding breath.*—Explain the correct manner of holding the breath and have the class practice it a few times. Explain how the coach observes the pupils' breathing by watching their backs.

(6) *Position of thumb.*—May be either over the stock or on top of the stock, but never along the side of the stock. Explain why.

(7) *Joints of finger.*—Trigger may be pressed with first or second joint. Second joint preferable when it can be done conveniently.

(8) *Prone positions.*—(*a*) Demonstrate correct prone positions with and without sandbag rest, calling attention to the elements which go to make up a correct prone position—body at the correct angle, legs spread well apart, position of the butt on the shoulder, position of the hands on the rifle, position of cheek against the stock, and position of elbows.

(*b*) Mention the usual faults which occur in prone position.

(*c*) Demonstrate the correct position again.

(9) *Sandbag rest position.*—(*a*) Demonstrate in the same manner as described above for prone position.

(*b*) Demonstrate coach adjusting sandbag to the pupil.

(10) *Sitting position.*—Demonstrate in the same manner as described above for the prone position.

(11) *Kneeling position.*—Demonstrate in the same manner as described above for the prone position.

(12) *Today's work: position exercises.*—(*a*) Demonstrate the duties of a coach in a position exercise, calling attention to each item.

(*b*) Demonstrate the position of the coach. Always placed so that he can watch the pupil's finger and eye.

(*c*) Place a group on an elevated platform and show how the instructor organizes it by employing the coach-and-pupil method so as to keep every man occupied.

(*d*) Continue the long-range shot group work today.

(13) *Do not squeeze the trigger today.*—Take up the slack in these exercises but do not squeeze the trigger.

(14) *Keep the blank forms up to date.*—Examine each man in the unit or team at the end of the day's work and assign him a mark.

(15) Are there any questions?

(16) Next lecture will be _____. (State hour and place.)

■ 216. THIRD LECTURE: TRIGGER SQUEEZE.—*a.* The following equipment is necessary for the demonstration:

(1) One automatic rifle.

(2) One sandbag.

(3) One box with small aiming target.

b. The following subjects are the ones usually discussed in the third lecture.

(1) *Trigger squeeze important.*—Explain that in firing at stationary targets the expert shot learns to increase the pressure on his trigger only when the sights are in correct alinement on the target. When the sights become slightly out of alinement, he holds what he has with the finger and only continues the increase of pressure when the sights again become properly alined.

(2) *Sandbag rest.*—Explain why it is used in trigger-squeeze exercise.

(3) *Pulsations of body.*—The natural movements of the body and its pulsations produce more or less parallel movement of the rifle. Very often men who are apparently very unsteady make good scores. You thus see that if you press the trigger the shot is displaced only by the amount of the parallel movement and will be a good one. But if you give the trigger a sudden jerk, you deflect one end of the rifle, and the shot will be a poor one.

(4) *Aim and hold.*—Any man can easily learn to hold a good aim for 15 to 20 seconds which is a much longer period than is necessary to fire a well aimed shot. Poor shots are usually the men who spoil their aim when they fire the rifles.

(5) *Calling shot.*—Explain calling the shot and why it is done.

(6) *Today's work: trigger-squeeze exercise.*—(*a*) Demonstrate the duties of a coach in the trigger-squeeze exercise by calling attention to each item.

(*b*) The work is carried on as in position exercises, with the pressing of the trigger added.

(*c*) Practice in the prone position only this morning, first with, then without, the sandbag.

(*d*) Finish up the making of long-range shot groups today.

(7) *Keep blank form up to date.*—Examine each man in the unit or team at the end of the day's work and assign him a mark.

(8) *Final word.*—Do not let yourselves become bored with this work. It is easy to learn, but it takes a lot of practice to train the members and to get in the habit of doing the right thing without thinking.

(9) Are there any questions?

(10) Next lecture will be _____. (State hour and place.)

■ 217. FOURTH LECTURE: RAPID FIRE.—*a*. The following equipment is necessary for the demonstration:

(1) One automatic rifle.

(2) Two magazines.

b. The following subjects are the ones usually discussed in the fourth lecture:

(1) Trigger squeeze the same as in slow fire.

(2) *Meaning of rapid fire.*—Rapid fire is nearly continuous fire. Rapidity comes from skill acquired by practice.

(3) *Keeping eye on target.*—Explain the advantages of this and how it gains time.

(4) Stress importance of aiming and trigger squeeze.

(5) *Application in war.*—Explain the advantages of keeping the eye on the target in combat.

(6) *Changing magazines.*—(*a*) Explain how this is done.

(*b*) Demonstrate it.

(7) *Today's work: changing magazines.*—(*a*) Explain how exercises are to be carried on.

(*b*) Demonstrate them.

(8) *Keep blank forms up to date.*—Examine each man in the unit or team at the end of the day's work and assign him a mark.

(9) Are there any questions?

(10) Next lecture will be _____. (State hour and place.)

■ 218. FIFTH LECTURE: EFFECT OF WIND AND LIGHT; SIGHT CHANGES; SCORE BOOK.—*a*. This part of the preparatory instruction can be given on any day in which the weather forces the work to be done indoors. If no bad weather occurs, this work should follow rapid fire instruction.

b. The following equipment is necessary for the demonstration:

(1) One D target. This target to be mounted on a frame and marked with the proper windage and elevation lines.

(2) Eight spotters that can readily be stuck into the target.

(3) Each man to have his rifle and a score book.

c. The following subjects are the ones usually discussed in the fifth lecture:

(1) *Targets.*—(*a*) Explain the divisions on the target and give the dimensions of each.

(*b*) Call attention to elevation lines. Have class compare them with diagram in the score book. Explain why lines are farther apart as the range increases.

(2) *Weather conditions.*—All weather conditions disregarded except wind.

(3) *Wind.*—(*a*) Explain how the direction of the wind is described.

(*b*) Explain how the velocity of the wind is estimated.

(*c*) Explain the effect of wind. Effect increases with distance from target.

(4) *Windage for first shot.*—Show the Table of Wind Allowances shown in paragraph 8*c* on page 9 of W. D., A. G. O. Form No. 82, and explain its use.

(5) *Elevation rule.*—State rule and explain it.

(6) *Light.*—Explain effect.

(7) *Score book.*—(*a*) Explain the uses of score book on range.

(*b*) Have class open score books and explain items of keeping score.

(8) *Exercises.*—Give the class a number of small problems as a demonstration as to how the day's work is to be carried on.

(9) *Today's work.*—(*a*) Study and practice in sight setting, sight changing, and the use of score book. Leaders and instructors will work up problems for their units or teams.

(*b*) Additional practice in the exercises of the preceding days.

(10) Are there any questions?

(11) Next lecture will be _____. (State hour and place.)

■ 219. SIXTH LECTURE: RANGE PRACTICE.—This lecture and demonstration should immediately precede range firing. If the class is not too large, it should be given on a firing point of the rifle range.

a. The following equipment is necessary for the demonstration:

(1) Material for blackening sight.

(2) One automatic rifle with gun sling.

(3) One sandbag.

(4) Corrugated type dummy cartridges (see par. 18).

b. The following subjects are the ones usually discussed in the sixth lecture.

(1) *Preparatory work applied.*—Range practice is carried on practically the same as the preparatory exercises except that ball cartridges are used.

(2) *Coaching.*—Coach watches the man, not the target. Coach does not keep the score for the pupil. Pupil must make his own entries in his score book. Coach sees that he does this.

(3) *Officers and noncommissioned officers.*—(*a*) Supervise and prompt the men acting as coaches.

(*b*) Personally coach pupils who are having difficulty in making good scores.

(4) *Spotters.*—(*a*) Use in both slow and rapid fire.

(*b*) If a spotter near the edge of the silhouette bothers the pupil in aiming, it may be removed before he fires again.

(5) *Watching eye.*—Explain how this indicates whether or not the pupil is squeezing the trigger properly.

(6) *Position of coach.*—Demonstrate in each one of the positions.

(7) *Demonstration of coaching in slow fire.*—(*a*) Place a man on the firing point and show just what a coach does by calling attention to each item.

(*b*) Demonstrate the use of dummy cartridges in slow fire.

(8) Demonstration of coaching in rapid fire.

(9) Read safety precautions.

Section IV

MARKSMANSHIP—AIR TARGETS

■ 220. Preliminary Preparation.—*a.* The officer in charge of automatic rifle antiaircraft training should be thoroughly familiar with the subject, should have detailed at least three officers as assistant instructors, and should train the assistant instructors and a demonstration group before the first training period.

b. He should inspect the range and equipment in sufficient time prior to the first training period to permit correction of deficiencies.

■ 221. Description of Miniature Range.—*a. Horizontal target.*—This target is designed to represent a sleeve-target towed by an airplane flying parallel to the firing point.

b. Double-diving and climbing target.—This target is in two sections. The right section is designed to represent a sleeve-target towed so as to pass obliquely across the front of the firing line in the manner of an airplane diving, if run from left to right, or climbing, if run from right to left. The left section is the same but represents an airplane diving from right to left and climbing from left to right.

c. Overhead target.—This target is designed to represent a sleeve-target towed by an airplane which is approaching the firing line and will pass overhead, or when run in the opposite direction represents an airplane that has passed over the firing line from the rear.

d. Size and speed of silhouette.—The black silhouette is a representation at 500 inches of a 15-foot sleeve at a range of 330 yards. It is 7.5 inches long. The speed of the silhouette should be between 15 and 20 feet per second. This speed represents that of an airplane flying between 150 and 200 miles per hour at a range of 200 yards. The size and speed of the silhouette are based upon the time of flight of the caliber .22 bullet for 500 inches. This time of flight is approximately 0.04 second. When the target is moving at a speed of 15 feet or 180 inches per second it will move 180×0.04 or 7.2 inches. Therefore, in order to hit the silhouette the aim must be directed approximately one silhouette length in front of it. If two or three target-length (silhouette-length) leads are used, the shot will hit in the appropriate scoring spaces. This does not hold equally true on the overhead target. If the shot is fired when the range is less than 500 inches from the firer, the lead necessary will be less than one target length.

■ 222. PREPARATORY EXERCISES.—*a.* A method of conducting the preparatory exercises is given in paragraph 139.

b. Each assistant instructor is assigned a target and conducts the preparatory training and firing of all groups on his target.

c. In preparatory training coach and pupil should change places frequently.

d. Forty-five minutes at each type of target should be sufficient to train each soldier in the preparatory exercises.

e. A detail of one noncommissioned officer and four or six men should be provided to operate each type of target.

■ 223. MINIATURE RANGE FIRING.—*a. Caliber .22 rifle.*—(1) The rifle should have the open sight.

(2) Two magazines for each caliber .22 rifle should be provided.

(3) Ammunition should be available immediately in rear of the firing line at each type of target.

(4) Coaches should load magazines as they become empty.

(5) Scorers should be detailed for each type of target. After each score is fired, they score the target. They call off the number of hits made on each silhouette and pencil the shot holes. The coaches enter the scores on the firer's score card.

(6) A platform permitting the scorer to score the target should be provided for each type of target.

(7) To stimulate interest, the instruction can be concluded with a competition between individuals, squads, or training groups.

(8) Targets as shown on figure 63 may be used on non-overhead targets for group firing of competitions. Only one target-length lead may be used in firing on this target.

(9) Considerable supervision is required in order to maintain target operation at the proper speed. This speed is

necessary because the lead is based upon speed of from 15 to 20 feet per second.

(10) Safety precautions must be constantly observed.

(11) Preparatory exercises with the caliber .22 rifle precede the firing of that weapon.

FIGURE 63.—Nonoverhead record firing target.

b. *Browning automatic rifle, caliber .30, M1918.*—If the size of the danger area permits, the Browning automatic rifle, caliber .30, M1918, is fired on the miniature range. Such firing is conducted in the same manner as with the caliber .22 rifle with the following exceptions:

(1) Sight over the top of the rear sight and front sight.

(2) The lead necessary to hit the black silhouette is approximately 2.5 inches. This is due to the difference in the time of flight of the caliber .30 and caliber .22 bullets for

500 inches. The time of flight of the caliber .30 bullet for 500 inches is 0.15 second. When the target is operated at the speed of 15 or 20 feet per second the silhouette will move approximately 2.5 inches during the time of flight of the bullet.

c. In sighting over the top of the rear sight and front sight the line of aim is lower than the trajectory of the bullet. Therefore it will be necessary to aim low in order to hit the silhouette.

d. Men must be constantly cautioned to keep the weight of the body forward. This is to prevent them from being pushed over by the recoil of the weapon.

e. Preparatory exercises with the Browning automatic rifle, caliber .30, M1918, precede the firing of that weapon.

▦ 224. TOWED-TARGET FIRING.—a. Range organization.—(1) Individual firing at a towed target being impracticable, all members of a rifle platoon, including both rifles and automatic rifles, are usually constituted as a group for such firing. A group the size of a platoon is the most convenient group for such firing.

(2) An ammunition line should be established 10 yards in rear of the firing line. Small tables at the rate of one per 10 men in a firing group are desirable.

(3) Immediately in rear of the ammunition line the ready line should be established.

(4) The first platoon or similar group to fire is deployed along the ready line with each individual in rear of his place on the firing line. Other platoons or similar groups are similarly deployed in a series of lines in rear of the first unit to fire.

(5) Upon command of the officer in charge the group on the ready line moves forward to the firing line, securing ammunition en route; other groups close up.

(6) Upon completion of firing by one group it moves off the firing line, passing around the flanks of the ready line so as not to interfere with the group moving forward.

(7) An ammunition detail sufficient to issue ammunition to groups as they move forward to the firing line and collect unfired ammunition from the group which just completed firing should be provided. These two operations should be

performed simultaneously. Unfired ammunition is delivered to the statistical officer.

(8) The officer in charge should have at least three assistants—two safety officers and one statistical officer.

b. *Ammunition.*—(1) Ball or tracer ammunition may be used. Tracer ammunition is useful to show the groups waiting to fire the size and density of the cone of fire delivered by the firing group.

(2) Tracer ammunition will assist the officer in charge in verifying the lead announced in the fire order. It also provides a means of checking the firer's estimate of the lead ordered.

c. *Technique of antiaircraft fire.*—(1) *Leads.*—The lead used in the technique of antiaircraft fire described in paragraph 136b is the average of two theoretical extremes. For example, if the maximum slant range to a passing airplane is 600 yards and the minimum slant range is 300 yards, the lead used would be that required for a slant range of 450 yards. Fire is delivered with one fixed lead in order to simplify the procedure. Experience indicates such a technique is readily taught and that it is effective.

The lead table given below may be helpful. It is based upon a 15-foot sleeve towed at 200 miles per hour and caliber .30, M2, ammunition.

Slant range	Lead required
100	2
200	5
300	8
400	11
500	14
600	18

(2) *Fire distribution.*—The usual technique of fire is described in paragraph 136. If time and ammunition allowances permit other methods may also be taught.

(3) *Variable lead method.*—(a) In this method the automatic rifleman fires each shot with a different lead. The maximum lead is used when the target enters and again when it leaves the firing area. The minimum lead is used

when the target is directly opposite the firing line.   Example: If three rounds are to be fired as the sleeve target passes across the front of the firing line, the first round is fired shortly after the target enters the firing area, the second round is fired when the target is near the center of the firing area, and the third shot is fired shortly before the sleeve leaves the firing area.   The fire order given by the officer in charge is: 1. SLEEVE-TARGET APPROACHING FROM THE LEFT (RIGHT), 2. LOAD, 3. 14–8–14 TARGET LENGTH LEADS, 4. THREE ROUNDS, 5. COMMENCE FIRING.   In this example it is expected that the three shots will be fired at slant ranges of approximately 500 yards, 300 yards, and 500 yards, respectively.

(b) This method has given good results but is more difficult to apply than the prescribed method.

(4) *Safety precautions.*—Safety precautions as given in paragraph 149 must be rigidly enforced.   This requires constant supervision by the officer in charge.

d. The results of all towed-target firing should be recorded and analyzed.   The statistical officer should record the total number of rounds fired and the hits obtained on each target. If the number of hits falls below the number expected, the reason should be sought and explained to the men.   On the other hand, when results are satisfactory the men should be impressed with the value of rifle antiaircraft fire.

e. *Estimating ranges.*—Training in estimating ranges of air targets is conducted by having individuals observe airplanes flying at known ranges.   The individual bases his estimate on the appearance of the airplane at key ranges. The following table, based on an O–46A airplane (observation, 1936), will be useful:

Range (yards)	Details seen
1,000	General outline of airplane.
700	Wheels, rudder, wing struts, and tail skid.
500	Antenna and small projections from fuselage.
200	Symbols and numbers. Letters on airplane can be seen plainly.

A flying mission of from 1 to 2 hours is sufficient for the instruction of a large group in estimating ranges.

SECTION V

TECHNIQUE OF FIRE

■ 225. GENERAL.—The instructor should secure necessary equipment, inspect ranges, and detail and train necessary assistants, including demonstration units, prior to the first period of instruction. Instructors should use their initiative in arranging additional exercises in the application of the principles herein contained. It should be explained to trainees how the exercises used illustrate the principles in the technique of fire. Good work in the conduct of the exercises as well as errors should be called to the attention of all trainees.

■ 226. RANGE ESTIMATION.—*a*. A number of ranges to prominent points on the terrain should be measured so that a few minutes of each period can be devoted to range estimation.

b. Range cards as shown below will be of assistance in figuring percentage of errors.

RANGE ESTIMATION

Name _____

Company _____

Squad _____

Number	Estimate	Correct	Percent	Remarks	Number	Estimate	Correct	Percent	Remarks
1					21				
2					22				
3					23				
4					24				
5					25				
6					26				
7					27				
8					28				
9					29				
10					30				
11					31				
12					32				
13					33				
14					34				
15					35				
16					36				
17					37				
18					38				
19					39				
20					40				

(Front)

TABLE FOR COMPUTING ERRORS IN RANGE ESTIMATION

Range (Yards)	Error (yards)										
	5	10	15	20	25	30	35	40	45	50	100
250	2	4	6	8	10	12	14	16	18	20	40
275	2	4	5	8	9	11	13	15	16	18	36
300	2	3	5	7	8	10	12	13	15	17	33
330	2	3	5	6	8	9	11	12	14	15	30
350	1	3	4	6	7	9	10	11	13	14	29
380	1	3	4	5	7	8	9	11	12	13	26
400	1	3	4	5	6	8	9	10	11	13	25
420	1	2	4	5	6	7	8	10	11	12	24
440	1	2	3	4	6	7	8	9	10	11	23
460	1	2	3	4	5	7	8	9	10	11	22
480	1	2	3	4	5	6	7	8	9	10	21
500	1	2	3	4	5	6	7	8	9	10	20
520	1	2	3	4	5	6	7	8	9	10	19
540	1	2	3	4	5	6	7	8	9	10	19
560	1	2	3	4	4	5	6	7	8	9	18
580	1	2	3	3	4	5	6	7	8	9	17
600	1	2	3	3	4	5	6	7	8	8	17
620	1	2	2	3	4	5	5	6	7	8	16
640	1	2	2	3	4	5	5	6	7	8	16
660	1	2	2	3	4	5	5	6	7	8	15
680	1	1	2	3	4	4	5	6	7	8	15
700	1	1	2	3	3	4	5	6	6	7	14
720	1	1	2	3	3	4	5	6	6	7	14
740	1	1	2	3	3	4	5	6	6	7	14
760	0	1	2	3	3	4	5	5	6	7	13
780	0	1	2	3	3	4	4	5	6	6	13
800	0	1	2	3	3	4	4	5	6	6	13
850	0	1	2	2	3	3	4	5	5	6	12
900	0	1	2	2	3	3	4	4	5	6	11
950	0	1	2	2	3	3	4	4	5	5	11
1,000	0	1	2	2	3	3	4	4	5	5	10

NOTE.—Example of the use of this table: Suppose the correct range to be 695 yards and the estimated range to be 635. The "error in estimate" is consequently 60 yards. Select two "errors in estimate" in the 700-yard space (the nearest to the correct range given in the table) whose sum is 60 yards, as 50 and 10. Add the percentages shown thereunder, and the result will be approximately your error. In this case:

7 plus 1 = 8 percent.

(Rear)

■ 227. TARGET DESIGNATION.—The time devoted to target designation should include careful instruction in target designation by each of the following three usual methods:

a. Tracer bullets.

b. Pointing.

c. Oral description.

While more time is required to teach oral description, it must be impressed on the men that all of the methods are important and have their application. Instruction is preferably conducted on varied terrain.

■ 228. RIFLE FIRE AND ITS EFFECT.—This step in instruction can best be covered by the use of a blackboard and several automatic riflemen firing tracer bullets to demonstrate the trajectory, danger space, dispersion, and classes of fire.

■ 229. APPLICATION OF FIRE.—a. Sufficient time and explanation should be devoted to the method of fire distribution to insure that all men fully understand it and can explain it in their own words.

b. A demonstration group simulating firing should suffice to show the technique employed in assault fire.

■ 230. LANDSCAPE-TARGET FIRING.—a. An explanation and demonstration will be necessary to show the technique and procedure of zeroing rifles and the firing of exercises on the landscape targets.

b. Units should be given practical work in writing fire orders for targets on the landscape panels prior to their firing any exercises.

■ 231. DISTRIBUTED FIRE.—Instruction in distributed fire required in field firing may be given the soldier by utilizing the rows of silhouette targets on the 1,000-inch target, U. S. rifle, caliber .30, M1, prior to instruction in firing at field targets. See figure 39.

■ 232. ASSAULT FIRE.—a. After the completion of record practice, and subject to authorized ammunition allowances, all men who have completed record practice, with the exception of antiaircraft troops, may fire the following table:

Assault fire

Range (yards)	Time (seconds)	Shots	Target	Position	Remarks
100_____	No limit.	5	Assault fire (see *b* below).	Assault fire___	Two magazines of 10 rounds each. (Fire while steadily advancing.)
100 to 125__	60_____	20	_____do_____	_____do_____	

b. The target is a screen 10 feet long and 3 feet high with three prone silhouette targets placed 1 yard apart directly in front of the screen. After the firing, each soldier marches up to the target to examine the effects of his firing.

■ 233. FIRING AT FIELD TARGETS.—*a.* The most difficult factor in the preparation of problems for field firing is the selection of the terrain which complies with the safety regulations contained in AR 750–10. A drawing should be made on a map showing all safety angles, target positions, and other required data.

b. The appearance of the ordinary prone or kneeling silhouette (E or F target) depends a great deal upon the direction of the sun, the background of the targets, and the angle at which the targets are placed. The effect of solidity can be obtained by using two figures placed at right angles to one another. The effect of fire distribution on a linear target can be determined by using a screen of E targets nailed end to end. The screen should be located so as not to disclose the position of concealed targets.

c. Maximum use should be made of the available terrain to permit the firing of as many units or teams from one firing position at one time as is possible. This firing should be controlled from a central location. Telephone communication between the firing point and the pits will facilitate this instruction. During this type of training, individuals and units should approach and occupy their firing positions with due regard to cover and concealment, after which men are rearrranged on the firing position according to the requirements of safety.

d. When sufficient time and ammunition are available platoon exercises should be conducted.

e. About 60 to 70 percent of the score allotted for the grading of units should be given for such parts of the exercise as the approach march and occupation of the firing position, fire orders, time required to open fire, rate of fire, and fire control. The remaining 30 or 40 percent should be given for the number of hits on the target and the number of target hits.

f. A 13-week training schedule should include about 24 hours for this instruction.

INDEX

	Paragraphs	Pages
Accessories	40	65
Actions, immediate	37	59
Advice to instructors	203–223	192
Aiming:		
Methods	127	130
Place in training	128	130
Aiming and leading exercises	141	139
Aiming and sighting, lecturing on	214	199
Air targets:		
Classification	132	134
For automatic rifle fire	131	134
Ammunition	41, 105, 210	67, 120, 194
Ballistic data	48	70
Care, handling, and preservation	46	69
Classification	42	67
Grade	44	68
Identification	45	68
Lot number	43	68
Storage	47	70
Antiaircraft—		
Fire:		
Leads for	134	135
Technique	133	134
Individual	136	135
Marksmanship training	138–142	136
Application of fire	181, 229	177, 215
Assembly and disassembly of rifle	4–12	4
Assault fire	183, 232	178, 215
Ballistic data	48	70
Beaten zone	177	174
Buffer, functioning	22	49
Care and cleaning of rifle	13, 14, 106	36, 37, 120
Care, handling, and preservation of ammunition	46	69
Cartridges:		
Dummy, use	18, 27	39, 55
Unfired, in rapid fire	117	121
Change lever setting	24	52
Change lever control setting	32	57
Coaches, duties	57	75
Coaching:		
Range practice	83	107
Prohibited	101	119
Concentrated fire	182	177
Courses fired	148	147
Critique of exercises	201	189
Cycle, functioning	20	42
Description	21	42
Delivery of fire	137	135
Demonstration of trajectories	180	176

219

	Paragraphs	Pages
Demonstrations, instructional	213	198
Disassembling rifle	8	4
Dispersion	175	174
Distributed fire	182, 231	177, 215
Duties of leaders	188	180
Elevation rule	71	95
Equipment	208	194
Exercises	196, 202	185, 190
Aiming and leading	141	139
Critique	201	189
Firing, situations for	200	189
Marksmanship training	58–69	75
Position	61–67, 140	80, 139
Preparatory	139, 222	137, 208
Rapid fire	69	92
Sequence	91	113
Sighting and aiming	58–60	75
Target designation	173	171
Trigger squeeze	68, 142	91, 140
Extractor, removal and replacement without disassembling rifle	11	31
Fire:		
Antiaircraft:		
Leads for	134	135
Technique	133	134
Individual	136	135
Application of	181, 229	177, 215
Assault	183, 232	178, 215
Classes	178	176
Concentrated and distributed	182, 231	177, 215
Control	186	178
Delivery	137	135
Discipline	185	178
Effect	179, 228	176, 215
Orders	90, 187	113, 179
Rapid:		
Lecture on	217	204
Procedure	99	117
Unfired cartridges in	117	121
Rate of	184	178
Rifle:		
Effect	228	215
Importance	159	162
Scope of instruction in	160	162
Slow:		
Procedure	98	116
Score, interrupted	111	120
Technique	125, 158, 225	129, 162, 213
Place in training	126	130
Firer, shelter for	103	119

	Paragraphs	Pages
Firing:		
Exercises, situations for	200	189
Field target	233	216
General considerations	198	187
Safety precautions	199	188
Scope of training	197	187
Group	146	146
Instructional	145	144
Landscape-target	230	215
Scope and importance	189	180
Weapons used	191	181
Line, organization	95	115
Miniature range	223	208
On wrong target	114	121
Pin, removal without disassembling rifle	10	29
Points	82	107
Procedure	150, 194	149, 184
Rifle	31	56
Care and cleaning, before, during, and after	14	37
Sequence	77	100
Towed-target	147, 224	147, 210
Fouling shots	109	120
Functioning cycle, rifle	20	42
Description	21	42
Functioning of—		
Buffer	22	49
Rifle	16–24	39
Trigger mechanism	23	50
Gas adjustment	33	57
Gloves and pads	108	120
Group firing	146	146
Gun sling, use	107	120
Handling of ammunition	46	69
Immediate action	37	59
Inspection of rifles	209	194
Instruction firing	145	144
Instruction in—		
Disassembly and assembly	4, 5	4
Firing positions	61–67	80
Functioning of rifle	16, 17	39
Immediate action and stoppages	35, 36	58
Marksmanship	49–51	71
Preparatory	52–75	72
Operation of rifle	25, 26	55
Rifle fire, scope	160	162
Target designation	167	166
Instruction practice on range:		
1,000-inch	86	109
Known distance	87	110
Instructions to pilots for towing missions	156	154
Instructors, assistant	207	193
Instruments, use	102	119

	Paragraphs	Pages
Landscape target:		
Description	190	181
Firing	230	215
Scope and importance	189	180
Weapons used	191	181
Preparation	192	181
Leaders, duties	188	180
Leading exercises	141	139
Leads:		
Antiaircraft fire	134	135
Determination and application	124	128
Lectures	213	198
Fifth—effect of wind and light, sight changes, and score books	218	204
First—sighting and aiming	214	199
Fourth—rapid fire	217	204
Place of assembly for	206	193
Second—positions	215	201
Sixth—range practice	219	206
Third—trigger squeeze	216	203
Loading of rifle	29	55
Magazine:		
Disassembling and assembling	12	33
Loading	28	55
Marking, target	97	116
Marksmanship:		
Courses, fired	78	100
Fundamentals	50	71
Object of chapter on	49	71
Training	205	193
Antiaircraft	138	136
First step—position exercises	140	139
Preparatory exercises	139	137
Preliminary preparations	220	207
Second step—aiming and leading exercises	141	139
Third step—trigger squeeze exercises	142	140
Moving—		
Ground targets	122	128
Fundamentals	123	128
Personnel	127, 128	130
Targets and ranges	129, 130	130, 133
Vehicles	124–126	128
Phases of	51	71
Preparatory	52	72
Coaches, duties	57	75
Elevation rule	71	95
Equipment	54	72
Examinations	75	96
Exercises:		
Position	61	80
Assault fire	66	90
Conducting procedure	67	90
Kneeling	65	83

	Paragraphs	Pages
Marksmanship—Continued.		
Training—Continued.		
Preparatory—Continued.		
Exercises—Continued.		
Position—Continued.		
Prone, with—		
Hasty or loop sling	62	82
Sandbag rest	63	83
Sitting	64	83
Rapid fire	69	92
Sighting and aiming:		
First—sighting bar	58	75
Second — alining silhouette and sights	59	76
Third—making shot groups	60	79
Trigger squeeze	68	91
Instruction method	56	74
Leaders, duties	55	74
Score books	74	96
Sight setting	70	95
Windage	72	95
When taken up	53	72
Zero, explanation	73	96
Method of aiming	127	130
Place in training	128	130
Miniature range	153	151
Description	221	207
Firing	223	208
Practice	143	144
Safety precautions	144	144
Misses	112	120
Firing on wrong target counted as	114	121
Nomenclature of parts of rifle	7	4
Officers, range	152	150
Operation of rifle	25–34	55
Organization of work	211	194
Pads and gloves	108	120
Parts, spare	39	64
Pointing targets	170	167
Position stoppage set-ups	204	192
Positions, firing:		
Assault fire	66	90
Exercises on	61	80
Procedure in conducting	67	90
Kneeling	65	83
Lecturing on	215	201
Prone, with—		
Hasty or loop sling	62	82
Sandbag rest	63	83
Sitting	64	83
Preservation of ammunition	46	69
Preparatory exercises	222	208
Procedure of firing	150	149

	Paragraphs	Pages
Range—		
Estimation	161–164, 226	163, 213
By eye	165	164
By observation of fire	164	163
Importance	161	163
Methods	162	163
Use of tracer bullets in	163	163
Officer	152	150
Practice	76, 79–88	100, 105
Coaching	83	107
Firing points	82	107
Instructional—		
1,000-inch range	86	109
Known-distance range	87	110
Lecture on	213–219	198
Organization	80	106
Safety precautions	88	111
Scope and object	76	100
Use of sandbag rest	81	107
Precautions	130	133
Ranges	121	125
Instructional practice on:		
1,000-inch	86	109
Known-distance	87	110
Miniature	153	151
Description	221	207
Moving	129	130
Towed-target	154	153
Rapid fire:		
Exercises	69	92
Lecture on	217	204
Procedure	99	117
Unfired cartridges in	117	121
Rate of fire	184	178
Record practice:		
For course D, 1,000-inch range	119	121
Regulations governing	89–119	113
Restrictions, rifle	104	120
Rifle:		
Automatic, Browning, caliber .30, M1918 without bipod:		
Care and cleaning	13, 14	36, 37
Description	2	1
Disassembly and assembly:		
Assembling	9	23
Care exercised in	6	4
Disassembling	8	4
Extractor, without disassembling rifle	11	31
Firing pin, without disassembling rifle	10	29
Magazine	12	33
Training in:		
Organization for	5	4
When taken up	4	4
Firepower of	3	1
Firing	31	56

	Paragraphs	Pages
Rifle—Continued.		
Automatic, Browning, caliber .30, etc.—Con.		
Functioning_____	16–24	39
Buffer_____	22	49
Cycle_____	20	42
Description_____	21	42
Explanation _____	19	40
Instruction in:		
Use of dummy cartridges for_____	18	39
When taken up_____	17	39
Object of section on_____	16	39
Trigger mechanism _____	23	50
Loading_____	29	55
Nomenclature of parts_____	7	4
Object of manual on_____	1	1
Operation:		
Gas adjustment_____	33	57
Instruction:		
Object of section on_____	25	55
Use of dummy cartridges in_____	27	55
When taken up_____	26	55
Magazine_____	28	55
Safety precautions _____	34	58
Setting change lever control_____	32	57
Storage_____	15	38
Unloading_____	30	56
Fire:		
Importance of _____	159	162
Scope of instructions in_____	160	162
Fire and its effect_____	228	215
Inspection_____	209	194
Zeroing-in_____	193	183
Safety precautions_____	34, 88, 144, 149, 199	58, 111, 144, 148, 188
Sandbag rests, use_____	63, 81	83, 107
Schedules, model_____	212	196
Score—		
Book_____	74	96
Lecture on_____	218	204
Cards _____	96	115
Shots included in_____	113	121
Scoring_____	96, 151, 195	115, 150, 185
Setting change lever_____	24	52
Shelter for firer_____	103	119
Shot groups_____	60, 176	79, 174
Shots:		
Recording of:		
Cutting edge of silhouette or line_____	110	120
Two on same target_____	115	121
Warming, fouling, and sighting_____	109	120
Sight:		
Changes, lecture on_____	218	204
Front, how to aline_____	84	95
Setting_____	70	199

	Paragraphs	Pages
Sighting and aiming, lecture on	214	199
Sighting bar instruction	58	75
Sights and silhouette alinement	59	76
Signals	157	155
Sling, use	107	120
Slow fire:		
Procedure	98	116
Score, interrupted	111	120
Spare parts	39	64
Stoppages	38, 92	59, 113
Storage of rifle	15	38
Target—		
Designation	135, 166–173, 227	135, 166, 215
By oral description	171	167
By pointing	170	167
Exercises	173	171
Importance	166	166
Instruction	167	166
Methods	168, 172	166, 171
Use of tracer bullets in	169	166
Details	94	114
Marking	97	116
Men not to know firer	93	114
Targets	120	123
Air:		
Classification	132	134
For automatic rifle fire	131	134
Field, firing	233	216
Description	190	181
Landscape:		
Preparation	192	181
Moving	129	130
Towed	155	154
Withdrawing of, prematurely	116	121
Technique of fire	125, 158, 225	129, 162, 213
Place in training	126	130
Telephones, use	100	119
Towed-target—		
Firing	147, 224	147, 210
Safety precautions	149	148
Scoring	151	150
Range	154	153
Towing missions, instructions to pilots on	156	154
Tracer bullets, use	163, 169	163, 166
Trajectory	174	174
Demonstration	180	176
Trigger—		
Mechanism, functioning	23	50
Squeeze:		
Exercises	68, 142	91, 140
Lecture on	216	203
Unloading rifle	30	56

	Paragraphs	Pages
Use of—		
Instruments	102	119
Telephones	100	119
Warming shots	109	120
Wind and light, effect of, lecture on	218	204
Windage	72	95
Work, organization of, in preparatory training	211	194
Zero:		
Explanation	73	96
Of rifle, determination	85	109
Zeroing-in of rifles	193	183
Zones, beaten	177	174

○

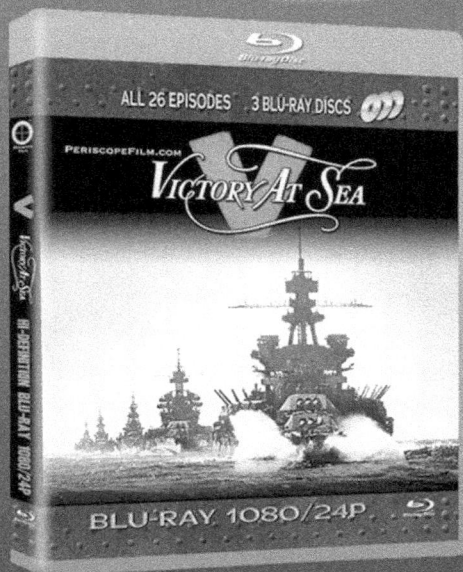
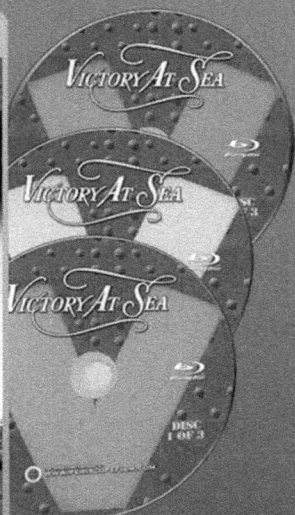
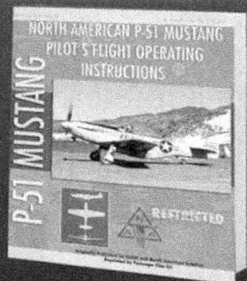

www.ingramcontent.com/pod-product-compliance
Lightning Source LLC
Chambersburg PA
CBHW052036090426
42739CB00010B/1928